D1017207

Toil

JODY PROCTER

Toil

Building Yourself

Chelsea Green Publishing Company

White River Junction, Vermont

Totnes, England

Designed by Christopher Kuntze.

Printed in the United States.
First printing, March 2000.

03 02 01 00   1 2 3 4 5

Printed on acid-free, recycled paper.

Library of Congress Cataloging-in-Publication Data
   Procter, Jody, 1943–1998.
     Toil : building yourself / Jody Procter
        p.   cm.
     ISBN 1-890132-67-5 (alk. paper)
        1. Procter, Jody, 1943-1998—Diaries. 2. Carpenters—United States—Diaries.
     3. Carpentry—Anecdotes. I. Title.
     TH140.P76 A3 2000
     694'092—dc21   99-058370

Green Books Ltd
Foxhole, Dartington
Totnes, Devon TQ9 6EB United Kingdom
44-1-803-863-843

Chelsea Green Publishing Company
Post Office Box 428
White River Junction, VT 05001
(800) 639-4099
www.chelseagreen.com

Go to your work and be strong, halting not in your ways . . .
—Rudyard Kipling, "The Song of the Dead"

Waking up this morning, I smile,
Twenty-four brand new hours are before me.
I vow to live fully in each moment
And to look at all beings with eyes of compassion.
—Tich Nhat Hanh, *Present Mind, Wonderful Moment*

Prologue

WHEN I started this diary in the fall of 1994, I was a few weeks shy of my fifty-first birthday. I was living in a medium-sized city in central Oregon with my wife, Kathleen, and our daughter, Shannon. We had recently moved to Oregon after what, for me, had been twenty-five years in California, my time split almost equally between San Francisco and Los Angeles. Like so many other restless, drug-fueled malcontents of my generation, I had fled the East Coast in the fall of 1968, arriving in San Francisco in my VW minivan with the Indian bedspread curtains and the FUCK COMMUNISM bumper sticker, in time to march down the Panhandle chanting "Free Huey!", stick daffodils in the chain-link fence at People's Park, and confront the devil at Altamont. With my long hair, love beads, and Fu Manchu mustache, I was another unwashed face in that giant counterculture army.

A year earlier, in the spring of 1967, I had graduated from Harvard *magna cum laude* with a major in Modern European History. For me, going to Harvard had been a kind of family obligation. The men in my family had been going to Harvard practically since the place opened, and money had never been an obstacle. As I drove west with my first wife, Star, I had $50,000 in my checking account, cash acquired by selling off the stocks in my trust fund. By hippie standards I was wildly rich. Star gave birth to our daughter, Emily, in 1969. And in 1972 we separated.

I spent the next twelve years in San Francisco as a performance artist, collaborating with my old college roommate, Taylor Moore, and a group of other friends in a collective guerrilla art project, whose highest expression came when we reenacted the Kennedy assassination in Dealey Plaza in Dallas in the summer of 1975. Though our work appeared extensively in museums and art galleries throughout North American, and even Europe, I never made anything more than bare subsistence money doing art. From the early seventies on, when my trust fund bank account ran dry, I supported myself as a carpenter.

In 1980, after Taylor Moore and I went through a seemingly irreconcilable falling-out, I abandoned my art career and moved to Los Angeles with my second wife, Kathleen. Shannon was born in December. A year later, at the end of a long run on drugs and alcohol, I joined AA. Life did not immediately become easy for me after that. I was depressed, unsure of myself, my identify as an artist was finished, I knew only a few people in L.A., and I was still a line carpenter, with neither the ambition nor the self-confidence to take my profession to the next level and go into contracting. Still, I enjoyed work in moderate amounts, and through most of the years of Shannon's childhood, I was able to work part-time and be a parent part-time; for me it was a perfect arrangement. Our decade in L.A. unfolded. As I remained sober and active in AA, I discovered an indistinct, non-sectarian spirituality slowly developing in me. Nothing dramatic, no burning bush, no blinding flash of holy light from the heavens—just a kind of mundane realization that who I was was more than what I did, that I had some kind of soul, that the unfolding of events was governed to some extent by what we call in AA a "Higher Power," and that even as my life on the outside continued to look kind of shabby and unaccomplished, my life on the inside continued to get better and better.

Sometime in the late eighties, still a carpenter but now living in Malibu in a trailer with an incredible view (we were really Malibu hillbillies), I realized that most of the time I was happy, very happy, sometimes even very, very happy. This was really something new for me. Despair and angst had always been the two central keys to my self-image. But now, in my mid-forties, I saw that good fortune had led me to a design/build company that was willing to keep me on with my part-time schedule; I was living in one of the most beautiful spots on Earth; I was solvent, healthy, and largely guilt-free. I feared nothing. I had become an avid long-distance runner. I loved my home life, my routines, my family. And even when I drove my old green '76 Toyota through the line-up of Mercedes and Land Rovers to pick up Shannon at the little Malibu elementary school, I figured myself to be one of the really lucky people in the world.

My parents died suddenly in 1989, one of cancer, one of liver failure—both of them dead within four months of each other. I spent a lot of time in hospitals in Boston that year, and was with them at the end, humbled and awed by this profound confrontation with mortality. In the months that followed, a modest inheritance allowed Kathleen and me to consider changing our lives a little, and one of the first things we decided was that maybe Malibu was not the best place on Earth for Shannon to go through her teenage years. She had just turned nine, and the materialistic pressure of living in one of the richest, most celebrity-choked small towns in the country was beginning to get to her. She was ashamed of her clothes, she was embarrassed to invite her friends home after school—friends in whose swimming pools our small trailer could easily have fit with room to spare.

So in the summer of 1991, after extensive research and discussion, we uprooted ourselves from our little Southern California paradise and moved to Oregon. A year later, fulfilling a long-held ambition of mine, using the last of the inheritance money, and with Kathleen's blessing, I enrolled as a graduate student in the creative writing department at a nearby university. Graduate school stirred up all my old and long-dormant artistic ambitions. During the two years I was a student, I wrote a number of stories, I wrote a novel, I received A's in all my classes, and for a while I managed to completely remove myself from my former reality, my reality as carpenter, husband, and father.

Reality for me became the library, the world of books and words and ideas. With my latent alcoholic capacity for total obsession, I became immersed and finally all but lost in this new realm. I wasn't drinking and I wasn't taking drugs, but I was, as Kathleen noted on more than one occasion, "somewhere else." Kathleen and Shannon, looking in on me from the outside, continued with their lives. Shannon was in school, Kathleen was a full-time clinical social worker, while I, if I was home at all, was back in my study, staring bleary-eyed into my computer screen, the floor and every other flat surface in the room thick with papers, open books, journals, and empty coffee cups.

It all ended with a graduation ceremony. My older daughter Emily and her boyfriend came down from Seattle. I marched in my cap and gown with the other M.F.A.s, I got my diploma, we went out for a celebratory luncheon, and then, suddenly, the two-year dream was over—I was back on the streets and I needed a job. Like some giant flywheel that keeps spinning and spinning and spinning long after the engine has been turned off, at first my brain would not slow down. Incapable of sustained and concentrated physical effort, I took on some small jobs: I re-did a bedroom for a friend, I built a small deck for an elderly couple, I made a retaining wall for a garden; all of which served me well, as manual therapy—but the money wasn't happening. Kathleen was getting a little nervous, and I knew that it was time to go out there and hook up with some company as a real, full-time carpenter.

September was upon us, winter was approaching, and the bills were piling up on the counter like early snow drifts. A friend of mine, with whom I play ice hockey on Sunday nights, was having a new house built for himself, and he knew I was a carpenter. He said the framing contractor was looking for an extra hand and asked if he could give the man my name. I said sure. One of the tenets of my slowly emerging spirituality is the belief that in order to be open to the miracles of the Universe you also have to remain open to whatever small gifts the Universe seems to offer spontaneously. This seemed like it might be one of those small gifts. Besides, I have an aversion to the job-search process. When a job is offered to me on a plate like this I have a very hard time saying no.

A few nights later, I got a call from Vern, the framer. He was looking for a carpenter, was I available? Yes, I said, I was. He wanted to know about my past experience, and I answered him with a long-winded monologue that probably left him wishing he hadn't asked. I described the last big residential project I had worked on, an all-concrete house in the Malibu Hills, built for an architect who hated right angles. The whole thing was laid out on a thirty/sixty-degree grid with tapered walls that in some places went from four feet thick at the base to nine inches thick twenty feet up. At the end of this unwieldy presentation—the rantings of a slightly desperate man—Vern said he'd be glad to hire me but he needed to know how much I wanted an hour. I told him that down in L.A. I made $15 per hour, but I understood things were a bit lower up here in Oregon. He allowed as how they were, and came back with an offer of $10. Never much of a haggler, I said okay, figuring I'd probably get a raise in a month or two, after he saw how skilled I was. The Universe had sent me Vern, so Vern's I was.

Off the phone, I went into the front of the house to tell Kathleen about my new job. Actually, I was kind of thrilled. Just to have a job at all after the dream-world of graduate school seemed like some kind of major accomplishment. And this was with a real hard-core Oregon construction company. I was looking forward to strapping on my belt and getting back out there among my fellow carpenters. The feel of a 24-ounce hammer in my hand, the smell of fresh-cut fir, the camaraderie and joy of working with good friends toward a common goal, the thrill of seeing a house go up.

I began this diary three months after I started working for Vern, on the day we broke ground for a new house. It seemed like a good day to start a diary. The entries carry through to the last day I worked on that particular house, the last day—as it turned out—that I worked for Vern, as well. The time period moves from November 1994 to June 1995, with one final postscript from a year later.

Toil

November 22

I T H A S B E E N so cold the past two days. I wear my insulated
boots with two pairs of socks, long johns, jeans, tee shirt,
turtle neck, Shetland sweater, jacket, and black stocking hat. And
gloves. Vern never wears gloves. Vern is the boss. I've been working for him for three months. Today we've been laying out the
foundation lines for the new house, a 3,000-square-foot house
for Beverly Chin along the third fairway of the Three Oaks golf
course. Because Brian, my co-worker, is off elk hunting in the
high desert, and Bud, Vern's son, is still up in the city with his sick
baby, it has been just Vern and me. Today we break ground. Spike,
the sixty-seven-year-old bulldozer driver, is two hours late because he had to go in for a prostate biopsy. The building is set
into the lot at an odd angle. The lot itself has only two sides that
are square, so it's hard to fix on anything. We operate off a bench-
mark out at the street, and Spike cuts the ground with the bulldozer blade, curling up rich, brown chunks of river-bottom dirt
as he outlines the footprint of the house. I stand most of the time
by the surveying instrument and indicate to Vern with hand signals where the pad has to be cut deeper. Most of the time I just
stand waiting for something to do. Freezing. We don't know yet
whether we will build the whole Beverly Chin house. So far, Vern

has the contract for the foundation—no more. There's no power out here yet, but the portable toilet did arrive today: boxy, plastic, pale green; it makes the place look like a real job site.

All day long, on and off, I keep thinking that my purpose in this lifetime is to learn humility. If that is true, then the more menial the job the better. I am fifty-one years old and a carpenter making $10 an hour in 1994. I made $10 an hour on a carpentry job in Hollywood in 1980. Is making low money a spiritual step? Time is the key. I watch the sun rise over the roofs of the other expensive homes that line the golf course. The air begins to warm.

At lunch-time I sit with Spike and Vern and listen to them talk about their trucks. Then Spike asks Vern how his father is doing. Vern's father is seventy-four. He quit doing carpentry a few years ago. "He stays home and keeps the fire going," Vern says. Vern is a third-generation builder. His grandfather was building homes here in town in the twenties. And then his father. And now Vern. And Bud, Vern's son, is the fourth generation. Picture Vern. I think of him as the old man, but he's only two years older than I am. Still, there's that careworn face, the slumped body. Six foot two, 175 pounds, with a bent-over frame and a face that has a vague suggestion of Don Knotts, but without the comic exaggeration. Maybe Don Knotts playing Gary Cooper in *High Noon*. Sometimes I think Vern hates building; there's a bitterness that seeps out in his sparse language. Even today, as Spike is hauling the dirt from the hole, dumping it into the trucks that come every half hour or so, Vern is muttering "on and on," as if he's just plain pissed off that everything always takes so long. The constant worries of the builder. At lunch, he and Spike trade stories about angry clients. I say, "On the whole I think people don't like construction workers." Vern agrees. "They're noisy, dirty, they smell bad." He swings his long arm around at the immaculate homes that line the street. The one directly opposite us looks like a place for Marcus Welby and his family. "They don't know what goes into making these places," Vern says, as if the dwellers of these suburban mansions were idiots, cut off from the true nature of their own dwellings.

Spike doesn't seem bitter at all. He's just bought that new John Deere bulldozer he's driving. And a new Ford truck, like Vern's, although he admits he's only put a thousand miles on it in six months. "I suppose, when I'm seventy-four," he says, thinking about Vern's father, back home tending the fire, "I may not have much ambition left." He's got to be sixty-seven or sixty-eight.

After lunch I go back to saying my mantra. I'm trying to be the old Zen ferryman. At a yoga retreat twenty years ago, I heard a swami from Oklahoma talking about work. "As a young man," he said, in his flat, Southwest drawl, "I had a job in a vacuum parts factory. I just kept saying my mantra, and I became the fastest, most efficient worker that vacuum parts factory had ever seen." Standing around is the hardest work. One more day until Thanksgiving, then four days off. When I'm out there in the cold I dream of things like hot-water bottles. We may not work tomorrow, as the forecast is for heavy rain. Vern says he will call later tonight. If we do work, we will start building forms, I guess.

I get home and sit paralyzed in front of the TV, drinking hot coffee and watching up-dates on the O.J. trial. Before I get into bed, I stare at myself in the mirror. This is me—a line carpenter for the Right-Wing Christian Construction Company. Is this it? Is this my dream come true?

November 28

BACK ON the job after a long Thanksgiving break. The air is warmer than last week, but with periods of rain. I wear insulated boots, long underwear, jeans, tee shirt, turtleneck, Shetland sweater, and my old bomber jacket and hat. My tool belt, held up with red suspenders on the outside of my jacket, is eight years old and some of the pouches are ripped out. All the small tool compartments are useless; pencils and nail-punches drop right through to the ground. Vern's belt is even older and funkier than mine, with a lot of pop-rivet repair evidence and several large areas covered with duct tape. For a while in the morning, I wear my thin, blue glove liners, but they get muddy so quickly I pack them away and work the rest of the day with bare hands.

Brian is back from his hunting trip to the high desert and spends the first hour of the morning telling me in endless detail how he and his Dad each shot an elk, totaling between them over half a ton of dead meat. He describes himself armpit deep in the elk carcass, gutting it out in the snow in the fading light in sub-zero temperatures in a tee shirt. Brian is nineteen years old, over six feet tall and more than two hundred pounds, newly married, a state champion high-school football player, and a marathon talker. He looks like Huck Finn grown into a jolly gentle giant, with a broad, goofy face and a gap-tooth smile. Vern putters around the site doing calculations, setting up work, and I listen to Brian babble on and on about the technicalities of the elk hunt. An "uh huh" or "really" now and then is enough to keep his jaw flapping. I often find his chatter irritating, as it constantly pulls my attention out of focus, away from the work.

We spend today lining out and setting up the forms for the footings of the house. The footings are six inches deep and fourteen inches wide. These will all be poured first and then the stem walls for the foundation will be poured a week later. The critical part of footing construction is the height. The foundation stem walls, which will be centered on the footings, are only six inches wide, so there is plenty of room to play with them, side to side. But the height is critical. As Vern says, you can always shim up but if the footing is too high we're in trouble.

Vern never uses bad language. I have seen him angry a few times; once when he was working a complicated roof out on the Baxter house (our last job), the numbers weren't coming out right, and Brian and I saw him throw his old, battered six-foot level at least ten feet down the roof. We looked at each other like little kids watching a parent flip out—half laughing, half panicky.

Vern's son, Bud, is still not back on the job. Bud's one-year-old daughter, Carol Anne, who had a tumor removed from her brainstem last week, is back up in the city for chemotherapy, and Bud went up there with her and his wife. There is still no satisfactory explanation—it's not cancer but they don't know what it is.

Brian ends the day telling me Little Johnny Foulmouth stories. Tim, my old foreman when I worked for Paul Murphy down in L.A. used to tell these same jokes. Brian tells one I first heard years ago: This teacher is asking the class for words beginning with the letters of the alphabet. She starts with *a*, then *b*, then *c*, and so on. Little Johnny Foulmouth shoots his hand up for each letter, but she dares not call on him. Finally, she gets to the letter *r*, and since she has called on everyone else, and since she can't think offhand of a bad word that begins with *r*, she finally caves in and calls on him. "Okay. Johnny," she says, "what word do you have?" "Rat," he says, "yeah, rat with a fuckin' dick ten inches long."

Tomorrow we work on the foundation footings again. Rain is predicted. Brian is going to bring his barbecue and cook elk steaks for lunch. At the end of the day, my hands are grimy with mud and I have that shivery, fingernails-on-the-blackboard feeling until after my shower and a heavy dose of lilac body lotion I find on Kathleen's side of the bathroom. I go off to my meditation class smelling like the inside of a little old lady's purse.

November 29

Tuesday. The day is overcast with regular periods of rain. The temperature is in the high forties to low fifties, and feels warm enough if you stay busy. The hardest part is working all day with wet hands. Even if you wear gloves, your hands are soaked in a couple of minutes. Anyway, it's impossible to pull nails out of your pouch with gloves on. We finish most of the footing forms today. Vern seems loose and accommodating with the straightness of things, although Brian finally comes up with a piece of crudely bent rebar that is rejected toward the end of the day.

This is Bud's first day back in weeks. Short and squat and dark, his whole body type is in sharp contrast to his father's. He has a dour, cheerless demeanor most of the time and the problems with his little girl have not helped. He seems happiest when he's in his "Proud to be an American" tee shirt, and listening to Rush Limbaugh. We've had a long vacation from Rush. Bud is back on the job site for the first time in almost a month, a month he has spent driving back and forth to Portland with his little daughter. I feel so bad for him. It's hard to know what to say. We talk a little at the end of the day, after Vern and Brian have driven off. The best I can do is ask him questions about the whole thing and tell him how much my heart goes out to him. I can't imagine anything worse than having a child in severe medical difficulties. A fund has been set up at the bank branch where his wife works and I want to send them some money, as they have been burdened with loss of wages, travel and living expenses, and on and on. The main portion of the medical bills is apparently covered by their insurance.

Brian does bring his barbecue today, but he can find only four wet matches and no charcoal lighter, so he spends most of the middle of the day running back and forth to the bed of his truck where he has the thing set up, tending a temperamental fire in the rain. Eventually he cooks, or more like smokes, three elk steaks down to almost inedible toughness. At about 2:30 in the afternoon we stand around—Vern, Brian, Bud, and me—our hands black with mud, chewing on this petrified elk steak and trading hunting stories. Since I have never been hunting in my life, I have no stories of my own and have to ask questions when they talk. I find out, among other things, that at two hundred yards you can still aim exactly at your target.

We work all day down in the mud, setting the formboards, pounding in grade stakes, backfilling on the bottom of the forms, and then tying off the rebar that has been lying in the mud. I feel wet and cold for eight hours. Through the hardest parts of the day, I keep repeating prayers and mantras to myself, and also thinking that at my age there are certain advantages to experiencing time as it creeps by slowly. The best part of the day is getting a hot cup of coffee at home and climbing into a hot shower. Tomorrow the forecast is for heavy, heavy rain and high winds.

Brian repeats many times a day how much he likes the rain. A real native. He talks all day. I think he finds the quiet disheartening and sometimes I feel bad for him, that he's so young and just starting out and stuck with these three older men who experience the dailiness of life as harsh and burdensome. Brian still has bright streaks of child-like carefree happiness that come bursting out of him and I wonder—how long? how long? His wife is expecting a baby in June.

I'm not sure what we will do tomorrow, since the forms and steel are nearly finished. What a fast job! There will have to be an inspection and then scheduling the concrete delivery and then the pour maybe Thursday. I should get rubber gloves for the pour; I seem to have developed an allergy to concrete. Three years ago, after a monster, all-day pour in Malibu when I wore rubber gloves but still managed to get concrete all over my hands, I developed a skin condition under two fingernails. During the next month, the lines of infection spread until both fingernails turned black and fell off. I heard of one man who worked in concrete and covered his hands with Vaseline and then white cotton gloves and then rubber gloves to combat the allergy. I'm not that bad yet. But I certainly don't want to lose any more fingernails. Those two fingers of mine were in Band-Aids for two months.

November 30

The day is wet and raining on and off. I wear full raingear to start, and then get too hot. I end up with my big yellow rainpants with the straps criss-crossed around my neck and no jacket, even when it drizzles. Vern leaves after the first few minutes to look at another job, and Bud and I work together. Brian has called in sick. We are putting up the last few boards on the bottoms of the forms where they are needed, backfilling with gravel, and then tying up all the steel so it is at least two inches off the ground. It's fun to be tying steel again; I haven't had tie wire and rebar in my hands since the job in Malibu.

Bud comments on how quiet it is without Brian, and I say that I think Brian talks pretty much all the time. If he's not telling you a hunting story or something about his family or his past, he will start in on the plots of old movies. Yesterday, when Vern was trying to take readings with the surveying instrument, Brian was going on and on, recounting the story line of a TV movie about a New York cop who's having trouble with his marriage. I tell Bud I have learned to fade it out, which is largely true. When Brian came back from his hunting expedition and told me that he shot an elk, I heard myself saying, "Tell me about it," and at the same time wondering—how can I actually be soliciting anything more out of his mouth? I told him a few days ago, after it had just gone on too long, "Shut your pie hole!" Brian thought that was a really funny thing to say. But it was ineffective—he kept up the chatter at the same pace.

We finish up about noon. Bud takes off, Vern comes back, and I sit in the cab of his truck with him and eat lunch and listen to the noon news on the radio plus the scanner for the fire department. Bud and Vern are both members of the Spencer Valley Volunteer Fire Department and are wired in to all emergency calls. We hear about a child choking on a piece of macaroni and a couple of motor vehicle accidents. Vern was up at 3:00 A.M. last night on a false alarm—a street light that looked to somebody like a house on fire. He is exhausted and dozes a bit as we sit in the warm truck cab and watch the rain pour down onto our muddy construction site.

After lunch, we cut eighty-five pieces of rebar that will be the uprights we stick into the wet concrete when it is poured. These in turn will tie to the footings of the stem wall. We finish with the rebar and call it a day. I get home about 2:00, take a shower, warm up, and fall asleep in bed after reading a few pages of Thomas Merton's *Asian Journals*.

December 1

This morning, we meet back at the Baxter house for a few final pick-up jobs so Vern can get his last check. We framed this 3,200-square-foot house in September, October, and half of November. I find out today that his bid for the framing was $28,000. That's just for labor. The materials, he thinks, would have been about $35,000. He says in the old days, labor and materials were about even, but no more. The price of wood goes up, up, up. No more trees left out there. And what's left is protected. This is a subject we stay away from. Ironic that I have such sympathy for the environmentalists and old-growth defenders such as Earth First! and am still a willing worker in an industry that uses lumber as its primary material. Even with the all-concrete house I worked on in Malibu, we used tons of lumber to build the forms—much of it carted away to landfill after it became too brittle and twisted to reuse. When I started working for Vern in September, he had framed up about half of the first-floor perimeter walls of the Baxter house. Total framing for the house, which was his contract, took about two and a half months, a fairly long time by industry standards, even for this house with its eleven gables.

Paul, the general contractor, meets us at the house at 8:30 to run through the punch list. Nothing too major, except one wall in the upstairs bathroom that is out of plumb. It's the mirror wall, and Paul is afraid it will show up when the mirror is hung. The client's wife will be looking at her shoes. No way to move it, I think; it's an exterior wall tied in to one of the main gables. Bud goes up later and beats on the base plate for about ten minutes with the sledge—nothing. The carpenter's ultimate solution to any problem—brute force. Don't know how that one will play out. We may have to cut reversed-taper firring strips.

I spend the day leveling window sills, cleaning up the exterior trim in a few places, installing a fold-down ladder for attic storage above the garage, nailing up drywall backing in a few places. The house is crawling with subcontractors. The client, Ray Baxter, my hockey-playing buddy (we play in the same over-forty league and it was through Ray that I heard about this job with Vern), has told everyone not to smoke because his wife is allergic to everything and now that the house is closed in he doesn't want any smoke toxins lingering. So everyone on the job is down on Ray. He can be a picky sort of character anyway, sometimes a little short on social tech. Through the whole job, while we were framing, he would come around and offer advice. Brian used to get so pissed off he looked like he was going to bury the claw of his hammer in Ray's big, shiny forehead.

Ray can even get to me. The last time it happened was when Ray was telling me to put a shim behind a window jam where I was just about to put a shim. I jumped down off my ladder and gave him a big bear hug and told him, "Ray, I just don't see how we would have ever built this house without your help!" He has kind of steered clear of me since then. Maybe he thinks I'm gay.

Today, while I'm installing the fold-down ladder in the garage, I'm listening to the heating subs ragging on Ray. They're both chain smokers, so his no-smoking ban hit them where they live. To compensate, they are both chewing wads of bubble gum and talking in weird accents; one in particular, a fat guy with his hat on sideways, keeps talking in this deep, monster accent and saying "the hell . . . !" and "what the fu . . . !" every time his partner says a word. They go into a long riff about another co-worker who plays "In A Gada Da Vida" on his steering wheel while he drives. "You got any more of that shit you're smokin'? Oh, I forgot . . . Ray."

I have lunch with Vern sitting in the main room, which is bristling with wires and light fixtures. "How you ever figure how much wire to use in a house like this I have no idea," Vern says, shaking his head. The electrician is stalking around, a kind of

Old Testament prophet with a beard, a lisp, a wild look in his eye, and a cellular phone. All the subs have cell phones on their hips and wander around, now outside with cigarettes, talking to godknowswho about godknowswhat. Vern has a cell phone in his truck, and he says it sure beats going down to the corner pay phone every ten minutes.

He leaves me alone in the afternoon. Bud has gone off to his wife's grandfather's funeral. At 3:00, Vern comes back to say that the inspector was over at the Chin job and said we had to wire in thirty more pieces of upright steel at critical spots. So back we go to the Chin project to cut and tie more rebar.

The day has turned fine, cool but dry with puffy clouds tinged with sunset pink. The rebar is cold and wet. My hands feel like they've been wet for a week. The skin is starting to crack around the thumb joints. Soon there will be big, red, chapped sores and I'll get extra sympathy at night when I come home. For now, I try to ignore the feeling of cold, wet, rusty, muddy half-inch rebar running through my fingers.

We finish up just after 4:30, which has always been our exact quitting time. Vern thanks me as I leave, a rare thing—maybe he's thanking me for staying an extra eight minutes. The concrete pour is scheduled for tomorrow at 1:30.

December 2

Pour day. I go out this morning and buy myself a pair of heavy-duty rubber gloves because of my concrete allergy. I already picture Brian teasing me about being a wuss. Real men move concrete around with their bare hands, he thinks. Until their fingernails start dropping off. Anyway, the pour never happens. I show up as scheduled at 12:30 and Vern is sitting in his truck looking like a whipped and pissed-off dog. Apparently a new inspector came by the job this morning and found still more problems, the most serious one having to do with a "mainte-nance easement" along the southeast boundary of the property. The house is designed to edge up right to the allowable five-foot limit to the property line, and someone down in the building de-partment has discovered this maintenance easement, which is ten feet. The easement is there to provide working access to a thirty-foot-high chain-link fence that protects the roof of Beverly Chin's parents' house, next door, from golf balls. The whole thing now has to be kicked back to the architect, the owner, the plan-ning department, the surveyors—who knows? Vern has sched-uled another pour for Monday, since this is all happening on Friday afternoon and there is no chance to resolve the dispute before the weekend. I leave work at 1:30 after tying in a few extra pieces of steel to the enlarged structural footing in the center of the garage walls. Vern seems genuinely apologetic, but what can he do? The building codes are a tangled, bureaucratic mess.

December 7

I WAKE UP this morning uncertain whether we will work or not—Vern has not called since Monday night and it is now Wednesday. The last time I talked to Vern, he said they were still "buttin' heads" down at the county building department over the maintenance easement. A call finally comes in about 8:30, just after Kathleen leaves to take Shannon to school.

The new plan is to pour the footings at 1:30 this afternoon. Vern has ordered the concrete, even though the footing is still not signed off by the inspector. Somehow, Vern's years and years of contracting give him the balls to challenge the inspectors this way. I have never been on a job where someone would think of pouring concrete before the forms were signed off. I get to the site about 1:00. The weather is overcast and cool, low forties, with a forecast for showers. Nothing like the snow we had the last few days in the foothills, and nowhere near as cold. Still, I wear long underwear, jeans, tee shirt, turtleneck, Shetland sweater, heavy jacket, and knit hat.

The pumper arrives at about 1:20, and we help him get his hoses set. The concrete comes in twenty minutes later. And twenty minutes after that, the inspector appears. We are already pouring. He is the third different inspector to show up on this project, and he says he doesn't know much about it. Doesn't even sign anything. Vern seems awfully laid back about these inspections.

The whole problem comes back to the Chin Seniors' thirty-foot golf-ball fence. For years, wild, sliced drives were busting the Spanish tiles on the Chins' roof, so they put up the fence to protect themselves. Our client, Beverly Chin, is the daughter of the Chin Seniors of the fence in question, and for now is living in their house. Apparently there has been a family meeting; they are going to make a reduction in the fence height and thereby circumvent the need for the ten-foot easement. Whether or not the new inspector knows anything about this future plan seems immaterial.

Meanwhile, we are pumping the uninspected footings full of concrete. We take turns with the hose, which is heavy enough to make the whole side of your body go numb in about fifteen minutes. A three-inch, semi-rigid rubber hose, held over the shoulder, it belches concrete and is controlled by the pump operator, a young kid in a camouflage hat and tennis shoes, who has a wireless switch attached to his belt. In the old days (five years ago), the pumper guy had a little switch on a wire that ran back to the pump. Or you had to wave hand signals at him, like a forefinger across the throat, if you wanted the pumper to stop pumping while you moved to a new location. Now it's a radio on a belt. High tech comes to the trades.

Vern follows behind us with the trowels. Watching Vern work is a study in swift certainty. He wields an old, worn-down trowel mindlessly, smoothing the surface of the wet concrete and flicking the little cumulative piles away from the sides, leaving an even, slightly rough surface, the perfect platform for the stem-wall forms. We pump sixteen yards, which is two full truckloads, and still come up about a quarter yard short. Again, on other pours, with other contractors of the past, there would have been a shit-fit about this shortfall. But Vern is cool. Just says we'll pour the last few feet of footing when we pour the stem wall, which is the second stage of the foundation.

No Brian today. He's come down with mono and will be out for at least two or three weeks. Apparently he got it during his elk hunt. Poor guy, his last day at work was the disastrous elk barbecue, and now he's down with mono, plus he has to move and he has no money for Christmas. Bud will be leaving next week to take his daughter up to the city for more chemotherapy, so it looks like it'll be me and Vern alone again, building the forms. Probably won't be that big a deal.

He tells me at the end of the day today to come in tomorrow and start knocking down the stakes. We'll leave the formboards in place until the concrete is a bit more cured—at least through the morning.

All in all, the pour goes well. We have no rain. In fact, it is actually sunny for part of the time, and otherwise we see varied clouds, with a beautiful sky, constantly changing. We have a gentle cloudburst at about 4:00, and a high thin rainbow appears against the dark eastern sky as the sun is already starting to set. Very green out on the golf course, but few golfers today. The forms with their streaks of concrete down the sides, their solid, screeded concrete leveled off on top, their fingers of rebar reaching toward the sky, feel full and complete. I'm used to concrete pours that are pure adrenaline charges with lots of shouting and anxiety and panic, but this one is so matter-of-fact it seems nearly over before it has begun.

Still no word on whether or not Vern will get the contract to build the entire Chin house. For now there is work ahead into next week—beyond that who knows? Even missing these last few days because of the permit screw-ups will take a big enough bite out of my check at the end of the month.

December 8

It's cold this morning. I wake up slowly. I was up late last night after the hockey game. If I play a late hockey game, it always takes two or three hours to get the adrenaline out of my system after I get home. I eat, I watch Letterman, I floss for half an hour. I must have been awake until 1:30 or 2:00. Several large black coffees (we're out of milk) and I am able to start moving.

I get to the site first and started knocking out all the stakes. Thin skins of ice lie across the old concrete puddles, but the day is dry, the sky clear, the stakes are dry. I wear double gloves: thin, blue polypropylene gloves for a first layer and my old, brown painters' gloves with the holes in the fingers for the second. They work okay. My hands stay warm and dry and my body starts to heat up after banging down twenty or thirty stakes. We have two sizes of the metal grade stakes, which are round, about three-quarter-inch thick, and either sixteen inches or two feet long, depending. Five hits with the single jack in one direction, then five in the other, and pull. Some of them are down deep into the mud and it takes all my double-handed, gloved strength to pull them out.

Vern shows up about 8:30 and starts going behind me, scraping off the sides of the forms and pulling the spreaders apart. The concrete has hardened, but it is still soft enough to scrape easily. We do not pull any formboards off until the afternoon. I pull stakes from 8:00 until 10:00, then scrape and clean up the top surface of the footings, then nail in the bottom spreader clips for the stem walls, which will be mostly two foot high and six or eight inches wide, depending. The clips get nailed off in the center of the footing and will hold each side of the three-quarter-inch plywood formboards. Bud has shown up by now, and I work with him on that while Vern is out ahead of us snapping lines.

Vern tells me at one point, "No matter how many times I

check it, there's always something off. I can't figure it out. I don't think I've ever set a foundation that didn't have something off." This is because one of the stem walls will have to be built slightly off-center on the footing, a factor that will not affect the house in any way. "Well," he says later, "you can only get it as close as you can, you can't get it perfect."

Around 11:00, I say something about how the sun is so pale and weak, it has hardly warmed up the air at all—we are still all puffing steam, though I take off my jacket when I start lugging the sheets of form plywood around. "Well," Vern says, "it could be a cold rain." And I say, "Hey, you don't hear me complainin'."

A good day for me, although I am worn out at the end of it and my arms ache. Vern says Connie, his wife, "will be trying to put the payroll together tonight." I say, "Good, I don't know how much longer I can hold off my creditors." And I say it joking, and he laughs.

I come home cold and muddy and mildly elated, I take a hot shower, and I stay in there a long time thinking what I really need is my own personal Jacuzzi.

Emily, my daughter from my first marriage, a love-child born in 1969, has invited us to dinner at her house. She has recently moved here from New York City, and I am ecstatic to have her living in the same town. Since her mother and I split up in 1972, we have mostly lived on opposite coasts. Through her childhood, I saw her only for vacations and summers. Now she lives practically around the corner. She's twenty-five and shares a place with her rock-and-roll boy friend who's out on the road with his band a lot of the time. She serves us pork chops and oatmeal-raisin cookies and later we watch *Seinfeld*. I sit in the rocking chair by the heater with the dogs and the family around and think how nice it is to work for a long day and to be bone-tired and cold and then to have the time off to eat and sit in a chair and feel warm. Short week. Tomorrow is Friday.

December 9

The day starts off cold but with no trace of rain. I have my double gloves on and they keep my hands warm all morning. By 11:00, the sun is out and the temperature rises to the low fifties. Vern has me clean all the concrete off the formboards. Four hundred feet of footings, probably a thousand feet or more of lumber, and I have to clean both sides plus the edges, so four sides altogether, or roughly four thousand feet of concrete-stained wood to be cleaned. My back is already tweaked from yesterday, and by about 9:30 I can feel it pull every time I lift one of the "pond-dried" two-by-sixes and flip it. Brian came up with this phrase—"pond-dried"—to illustrate the fact that these boards are wet and extremely heavy.

The cleaning of the lumber takes four hours. I do it with a flat-head shovel. As I work, I pray and try to set myself a rhythm, try to ignore the obsessive thought that I am the shit-kicker on this job, the grunt, the laborer when we need a laborer. On crews in Southern California, I never saw a non-Hispanic clean form lumber. And now I catch myself in this bitter, racist realization. I'm a white man—I shouldn't have to do this kind of shit. And I remember what my old friend and college roommate, Dan, said when I told him I had taken this construction job—"You have an amazing capacity to enjoy menial work"—and in some ways he knows me better than anyone, even Kathleen. But I want them to appreciate what it's like out here, I want everyone to marvel at my capacity for physical discomfort, at my proletarian leanings, I want them to understand what "back-breaking work" really means. For me, being a grunt carpenter is, at worst, an ego trip. A triumph in my long crawl to the bottom. Now I know what it's like to be cold and wet all day, and to have hands that are rubbed raw on splintery, dry wood and mud and concrete and rebar.

But today isn't like that. The air warms and the day becomes lyric, sweet, sunny. My mantra flows along just below the threshold of consciousness: "*om shanti . . . om shanti . . . om shanti . . .*" It's a mantra I took up years ago during a yoga retreat in Northern California. I don't even remember why I chose it. All I know is that *om* is the word for the divine totality and *shanti* is the word for peace. "*Om shanti . . . om shanti . . . om shanti . . .*" I say it over and over and over again to myself, accompanying the rhythm with the sound of the steady scraping of my flathead shovel against concrete-caked wood. The golfers come by, hacking away at their golf balls through the trees and I work and look up at them and smile with that careful and controlled contempt of the workingman on a work day looking on at men of leisure.

Bud is here today working but he will be gone most of next week, up to Portland again with his little daughter, who will have a small operation to have a port inserted into her heart, or one of the main veins, the same thing my mother had at the end for the nutrition IV, and then this little girl, Carol Anne, will begin chemotherapy again. They are told that there is no way to predict what the side effects will be, everyone responds differently.

At lunch, Vern and Bud tell stories about fires they have fought, including the time Bud single-handedly saved a whole Christmas tree farm. The heroics of this tale center around the spraying of one garden hose on one smoldering dead bush. Vern laughs as he must have laughed about it many times before.

After lunch, Vern's wife Connie shows up to do the payroll. She sits in her car for about half an hour crunching numbers, then comes out onto the site with the checks, or really only my check, and Vern calls me away from the form work I am doing—I have finished with the concrete cleaning—and introduces me. Connie is a short, sharp-eyed woman in her mid-fifties with gray hair pulled tight around her head and a bright, chatty manner—quite the contrast to Vern, who gives off words as if each one were a painful extraction. Among other things, Connie tells me how glad they all are to have me working for them. They had almost resorted to a temp agency, and, she says, they were "afraid they might end up with some drug addict or alcoholic." With a straight face, I tell them I understand how that could be a problem. They have no idea I am, myself, a drug addict/alcoholic, in recovery now almost thirteen years. Part of what I like about this job is its anonymity. They have no idea who I am and could care less.

The day ends on a positive note. I have my check for November's work in my pocket: $957.53. As always, a huge chunk has been taken out for Social Security, and I heard just a few days ago on the radio that they're planning to bounce the benefits age up to seventy. I could be out here slogging through the mud at age sixty-nine. And I guess I'd be happy to have the work. The Friday epiphany is somewhat muted by the fact that we have only worked a half week. Brian comes by late in the afternoon to pick up his check and he looks sickly and pale. Still, he says he's pretty much over his mono and he plans to come back to work on Monday.

December 12

BRIAN does show up this morning and it's good to see the kid up and about. I work with him all day, or at least until about 3:00, when he admits he's feeling little unsteady and weak. Vern tells him to go home and rest. Brian works well, although he is not really up to his usual superhuman standards of strength. At one point we are cutting off a piece of rebar that is sticking up too high, a few inches above the top of the stem-wall forms. He is holding it and I am cutting it with the Sawzall, which vibrates like crazy and has a hard time getting a real bite on the metal. It takes a few minutes, and when we are a bit past halfway, I tell him to snap it off. He can't and then I do, which has to be a first. In feats of strength he's usually ten times as strong as I am. But not today.

Vern seems particularly friendly and later in the day tells me he and Connie want to have me and Kathleen over for dinner with Brian and his wife and Bud and his wife—a regular, old-fashioned company Christmas party. I picture Vern's house like some big hunting lodge with cow-brand upholstery, elks' heads on the walls, and old antique construction equipment show-cased. Also, I'm immediately worried they'll be serving whiskey. Ahh well.

We finish up most of the stem-wall forms. We'll probably do the fine tuning tomorrow and then have to wait again for the building department and the variance on the maintenance easement. So there may be a few more days off coming up. If I had any money to spend on Christmas presents, I could use the time for Christmas shopping.

December 13

I arrive at the usual time and Vern is already there, parked in his truck and talking on the cell phone. I drag my tools out of the back of the car and we go to work. Only a few details to be punched out and the forms are ready to pour. I finish nailing off the rapi-clips, top and bottom. These are particularly important on the bottom, where the force of the concrete can cause the forms to float up. Still, compared to forms I have built in the past, these seem almost flimsy. The art of form carpentry seems to be this—to know exactly how little support you need to pour. My thing has always been overbuild, overbuild, when in doubt add a few more two-by-fours! Those sleepless nights worrying about a form blowout! But these aren't my forms, they're Vern's, so they're his to worry about. After the nailing, I go around the whole perimeter and nail off three-quarter-inch metal strapping to the outside corners. Again—a kind of minimal insurance. Vern is cutting in block-outs for the utilities: electrical, heating, plumbing. These are created by jamming pieces of pink four-inch Styrofoam down into the forms. We are finished with everything by 11:00 and wrap up the tools.

No Brian today; I don't ask if he has called in sick or if Vern has just called him and said "no work." At the end, when the tools are repacked in the truck, Vern tells me what he knows so far about the paperwork hassles that still need to be fixed before we can pour: the golf-ball fence, the maintenance easement along the east property line, a resurvey requirement, and an unknown time for the paper traffic to work its way from desk to desk in the city recorder's office. I tell him I'll just be on standby.

At the very end, I ask him how much variance in level he would expect over the entire top of the stem-wall forms (four hundred running feet, seventy-five to one hundred feet corner to corner). He says he hopes no more than a quarter inch. I say, "Aren't you going to check it with the instrument to find out?" He says, "At this point I don't want to know." And then he says, "Level gets lost pretty quick in the house anyway." I know just from seeing the way the forms sit that there must be irregularities bigger than a quarter inch, but he also says that he has sighted down the lines in a few places and is satisfied that overall the foundation is sitting level. The finely tuned eyeballs of a master builder! In the end, it seems he trusts his own intuitive sense of level as much as anything any instrument could ever tell him. If the forms are out of level by any significant degree, he seems to be telling me, then he would certainly know it. He would know the way a musician would instantly hear the loss of intonation in a complex symphony.

He gets into his truck and drives off. A wave. We'll be back to work whenever the boys in the city figure this one out. I'm off for a few days. I love the feeling—money or no money. It's out of my hands. I want to go and buy a latté and sit in a rainy café somewhere and relish the long moments of doing nothing.

December 21

WINTER Solstice. A delay of over a week. Vern called last night. "I was going to ask if you remembered me," he said. The pour of the stem walls is finally set for 3:00 this afternoon. The shortest day of the year. I get to the site at 2:00, the first one there. Tools out, my yellow rain pants on, not because it's raining, but for the concrete.

Vern drives up a few minutes later with the table for the radial arm saw, which he is moving over from the Baxter job. The last thing there, he hopes. I tell him it's a good sign, but Beverly Chin has still not formally committed to Vern as the general contractor for the house. "She's still trying to knock the price down and I'm trying to figure out where. But it looks good." He seems to think she'll come around. All the hassles with the golf-ball fence and the property line easements seem finally resolved. Vern tells me he had to go in person to the building department and claw his way up to the top levels before he could get any movement. The paperwork side of construction—I understand nothing about it; it seems like some kind of unspecified inner world, a network of old boys and connections. Still, people get building permits all the time, houses are built and remain standing. For me, the essence of construction is the play of the most simple processes: the hammers, the boards, the old power tools lying in the back of the truck. All these stamped pages and columns of figures drain my spirit. Give me something to nail, to cut, to clamp, to drill.

Brian shows up, having recovered from mono and then from a bout of chicken pox. Giant that he is, he does nothing in small doses. He seems to be back to his usual bouncy self, however. There is no immediate work to be done as we wait for the pumper and concrete trucks, so we sit on the sawhorses and he tells me in detail how he spent $1,000 on Christmas presents for his wife and family. He's a babbling phenomenon. When he runs to the bottom of his list of Christmas presents, he starts describing all the tricks he did as a kid on his Big Wheel on a hill in a park down the road from his house. Nineteen years old, I think, and already the details of his life could fill volumes.

Bud comes along, rubber boots and red plastic pants in hand, and tells a few firemen stories—like the time the tanker truck rolled down an embankment with all of them strapped into the cab. We are a bit awkward with each other because we have little experience just standing around. All our time together is spent either working or on formal breaks, but this afternoon we stand around killing time, waiting for the concrete to arrive. When it finally comes, it's "all asses and elbows" until well after dark.

This time the pumper is an old guy. He looks like the convict in the original *Great Expectations* movie, the same hook nose, the same gruff, threatening manner. He has to be in his late sixties at least, still slinging that big heavy hose around. He grumbles about how late it is, how much we have to pour. Two trucks' worth, about sixteen yards. But once the pumping starts there's no time to talk.

I work the hose with Brian. Brian, big and strong as he is, takes the front end and pumps the concrete down into the forms. I follow behind, lifting and moving the hose around all the obstacles. The three-inch, black rubber hose pulsates and purges the concrete with a spasmodic, almost peristaltic rhythm. I have a four-foot two-by-two for tamping and try to settle the concrete down into the forms, paying attention, particularly, that it flows well under the many eight-by-sixteen-inch ventilation ducts that are set near the top of the walls and will keep the crawl space dry. Twenty-six of these are scattered about. Bud follows behind, rough-screeding the concrete and making sure it reaches up everywhere to the top of the forms. Vern comes along last, finish-troweling and sticking in the anchor bolts, or "J" bolts as they are called because of their shape. They stick up out of the top of the wet concrete about two and a half inches and will hold the pressure-treated sill plates in place. The whole process takes a little over two hours. By the end, it is dark, and Bud pulls his truck into the driveway with its four big tracking spotlights above the cab, and we work in cold, foggy, black shadows, finishing up, hosing down the tools, packing everything away. It feels good to be back at work.

As we leave, Vern says he'll come out and check the concrete in the morning, but that he isn't sure it will be strong enough in less than twenty-four hours to start breaking down the forms. Tomorrow may be another day off. He says, also, that we will not work the day after Christmas. Brian tries to talk me into going with him to an antique store up in Junction City, a place where he spent part of his $1,000 Christmas present money.

December 22

Waiting at home this morning for Vern to call. He told me last night that he would call after he went out to check the concrete this morning. He calls about 9:30 to say he thinks it will be okay to work on the forms if we are real slow and careful at first. I get out to the site about 10:30, and Brian's big, rough-sounding Ford pickup pulls in a few minutes later. On the bumper, a bumper sticker says REAL MEN LOVE JESUS. Bud shows up about 1:00.

It gets so warm by mid-afternoon that I am working in just jeans and my Red Sox tee shirt. This must be the first time in months I have been peeled down to just a tee shirt. I even take my knit hat off. Brian calls it my "sponge," because I wear it sometimes in the rain and it soaks up the water. But now Brian has a sponge too.

We work the rest of the day breaking down the forms. Pulling the clips first, then scratching down the rough and uneven concrete to create a smooth surface for the sill plates. The concrete is still soft enough to scrape easily. In the afternoon, we pull the two-by-six that runs around the outside of the forms, the only side that will be seen, the only side that needs to be relatively straight. I'm surprised to see how much scalloping there is, but Vern seems unconcerned. In fact, in the end, after we have stripped all the forms and only concrete is showing, he says it looks like a pretty damn nice foundation.

Brian works away, talking, singing to himself, jabbering as he goes. Is he hyperactive or what? Is there an attention deficit disorder here? Did his mother drink heavily during her pregnancy with this chattering giant? Is there a pathology to this constant need to fill the silent void? He sometimes doesn't seem to care if I hear him or acknowledge him or anything. I spend several hours cleaning concrete off the plywood formboards, and because of the scraping sound, I can't understand a thing he is saying. He is trying, I think, to explain to me the details of some side job he has taken up—a tiered marketing scheme involving jewelry. Sounds like some kind of pyramid game, but there are big blank gaps in the explanation. I nod and smile and say "uh-huh" from time to time. I've told him several times that I'm deaf in one ear, but this doesn't seem to slow him down at all.

The day is fine. I feel tired in the end, but relaxed. A good pace to the work and a gorgeous blue sky, with sunshine and puffy, white clouds. The golfers are coming by regularly, their shots clicking in the background, and we look up and acknowledge them from time to time, particularly the ones that come in close to us. Next door a man spends a long time cleaning out the gutters of the Chin Seniors' house, and then tells us he has retrieved eight golf balls. And this is the house that is still protected by the thirty-foot fence.

Tomorrow we will put on the sill plates, then Christmas, then next week the floor joists, then it's time to wait again for the plumbers, the electricians, the insulators, and the always unpredictable inspectors. All that before we can sheet the floor and start really framing. It looks like a few more days off between Christmas and New Year's. I'm just as glad. But glad too that the $10 an hour I earn is not the only income for me and my family. We'd be out on the streets if it was. Kathleen has a good job working as a psychiatric social worker for a clinic here in town, so we've got enough.

December 23

The day before Christmas Eve. Another beautiful, warm day. In the sun, by midday we're almost hot.

Brian and I work together in the morning, cleaning and stacking the two-by-sixes, and then cleaning and packing away all the stakes and remaining debris from the pour. We establish a scrap pile near the street and pile all pieces of wood under two feet out there. It feels good to clean everything away from the walls of what is starting to seem vaguely like a house. Inside, on the gravel of the crawl space, the stem walls are two feet high and with their twenty-six eight-by-sixteen-inch ventilation shafts for windows, the place does begin to show characteristics of a primitive shelter.

The work is hard and steady but there is no getting away from the feeling of Christmas. Brian talks constantly about the presents he is expecting and the presents he is giving. He called me last night while I was at basketball practice (I'm coaching Shannon's eighth grade girls' basketball team) to ask me what Bud's wife's name is. He needed to write it out on a Christmas card. I got home too late to call him back and when I arrive this morning, he says he has already written on the card "Bud and family." He gives both Bud and Vern cards but then, during a slow-down in the work—waiting for a lumber delivery—he brings out a present for me. It's a thing called "The Mighty Boss," a blue, canvas tool organizer for my tool bucket. Since I came on the crew four months ago, I have always brought my belt and a bucket of other tools onto the job site with me. This "Mighty Boss" is an organizer with twenty or thirty pockets of various sizes that fits over the rim of the bucket, leaving the inside for the larger crowbars and hammers, but allowing me compartments for all the small stuff, the chisels and nail-sets and screwdrivers and bottles of chalk. Brian is so pleased with the present he seems about to burst. And indeed, I am touched he would think to buy me something so nice. His "workin' buddy" as he calls me. At the end of the day, we have another round of present exchanges. Vern doles out Christmas cards and $50 gift certificates to Fred Meyer (The Home of One-Stop Shopping) as Christmas presents. "I expect to see everyone with some new tools," he says. And I pass my presents around. A weird 1995 hand-drawn calendar Shannon and I made for Brian. Some fire starters I made for Vern. And for Bud, three cabinet doorknobs I made from golf balls that we found on the site when we were excavating. I have their presents all wrapped up and tell everyone not to open them until Christmas Day. I'm embarrassed to have them opened in front of me. Brian is unabashed in his delight at seeing people get their presents. He keeps saying he can't wait for Christmas so he can give everyone everything.

The work goes well in the afternoon and there is a camaraderie and sense of heightened expectation. We get most of the sill plates down and I build two of the five mid-span pony walls that will hold up the long TJI "Silent Floor" floor joists. Working on the pony walls is the first true woodwork framing I've done in nearly a month. I'm happy to be up out of the ground and back with straight-ahead carpentry. Cut, lay out, line up, toenail, end nail, eightpenny and sixteenpenny nails, drill for the anchor bolts, plate, and double top plate. Each wall has an enlarged opening for crawl access. The plumbers and telephone people and termite inspectors of the future will have a hole two feet wide and one foot high to squeeze through. Those people must have to be skinny as part of the job requirement.

We knock off at 4:30 and I head home, excited about the three-day Christmas weekend. I stop off at Toys "R" Us to join the holiday mayhem for a few minutes. It feels good, like just the right Christmas thing, to be lining up with the other last-minute shoppers. Even dirty and in boots and long underwear, I feel right at home as I inch my way down one of the long cash-register lines and present my credit card to pay for those last-minute stocking-stuffers. What a Dad!

December 27

BACK for a full day of work after the Christmas break. Brian pulls in and starts right off with a show-and-tell of his new tools, all the things he asked for on his Christmas list. He has some of them in his belt, but waits for the 10:00 break to bring out the rest to share. Cat's paw, small pry bar, three new hammers, a four-foot level, a Craftsman reciprocating saw, wrenches, a small toolbox to fit under the seat of his truck, a tri-square with a sixteen-inch ruler, which is unusual. Vern seems pleased that his gift certificate has been put to such quick use.

When I ask Vern about his holiday first thing this morning, he reports that Carol Anne, Bud's little girl, was back in the hospital Christmas Eve and Christmas Day and in fact is still there. Complications from the first chemotherapy treatment. Bud and Debbie (Bud's wife) have stayed at the hospital through the night. At least it's at St. Joseph's this time, here in town, and not up in the city.

The weather is warm and humid with a definite feel of impending rain. Brian and I finish up the three remaining pony walls by lunch. The work is good, clean and easy; we move fast. Brian recounts his entire Christmas experience, details menus of various meals, tells lame jokes, and when all else fails, sings a song that begins with the line "I know a song that gets on everybody's nerves

... it's the song that never ends ... I know a song that gets on every-body's nerves ... it's the song that never ends ..." Does he know this drives me crazy? A real piece of work, he is.

Bud shows up after lunch and the rain starts. I go off and put on my full rain outfit: yellow pants with cross straps, a dark green, heavy rubber jacket, and my "Gilligan" hat, the hat that looks like the one Gilligan was wearing during the storm that blew the castaways to their ill-fated island. It's a great hat, though, and it really keeps the rain out. This is the hat my Gloucester fishermen ancestors wore for generations. The air isn't particularly cold—mid- to high-forties. The only real dis-comfort is working continuously with wet hands. Wet hands that are soon caked with pressure-treated sawdust, concrete flakes, dark brown mud.

Still, I'm a nailing fool. I take a handful of eights, line them up with the heads all facing forward, and peel them out of the palm of my left hand one at a time with my thumb—peel, set, nail. I'm using my old 24-ounce wooden hammer with the duct tape around the neck, and in this soft wood I can sink those puppies in two or three hits—peel, set, tap, pound, pound, peel, set, tap, pound, pound—I set up a rhythm and move fast as we sheet the pony walls with half-inch wood composite, a new "forest prod-uct" somewhere between plywood and cardboard. Vern is laying out the joist hangers for the TJIs that arrive on a long flatbed truck about mid-afternoon. Some of these floor joists are forty feet long or more. These are a new thing in construction tech-nology. Still two-by-twelves, but made with fir two-by-twos that sandwich a thin piece of half-inch composite board. They weigh a third of what full fir two-by-twelves would weigh. And are sup-posed to be better in many ways—stronger, quieter, straighter. What do I know? They look like flimsy pieces of shit to me. But I love that they're so light weight. Brian can pick up a forty-footer all by himself.

At the end of the day, he drags Bud off to the back of his truck to show him all his new tools. "I've already seen 'em," I say.

"See you in the mornin'," Vern says. "Hopefully it won't be rainin'. But if it is . . ." He waits a beat as if to test me, then finishes the sentence: "too bad." He's joking of course, in his own, sometimes slightly sadistic way. Still, he's earned the privilege— he's been right out there in the rain with us all afternoon himself. I feel increasingly accepted by him. Accepted for what I am—a slow, steady $10-an-hour worker, who shows up every morning on time, who listens to everything that tumbles will-nilly out of Brian's mouth, who doesn't talk too much, who sticks to business, who understands most of the basic work that is going on, who is not one of those dreaded drug addict/alcoholics from the temp agency. It has been a kind of slow thaw between us. I'm clearly not his type exactly—not a hunter, not a Christian, not a Vietnam vet. He smells a trace of the old hippie in me, and of course, he's right. What he can't detect and never would detect, I suppose, is the Harvard graduate, the preppy from the old Boston family. And I'm grateful for that level of anonymity. To Vern I'm just another in a long line of carpenters who have walked onto his various job sites over the years. If I show up and do my wages' worth of work, he's happy.

We're going over to Vern's house tomorrow night for a company holiday get-together. Before I drive off, Brian slips me a bag with nine steaks from the elk he shot in November. What a guy. Now I feel like I should have gotten him a bigger present for Christmas. Oh well. It's a done deal. Wait'll next year.

December 28

We are scheduled to go to Vern's house tonight for the holiday office party, but his wife is sick and Carol Anne is still in the hospital, so he calls it off. I'm partly relieved, though in a sort of perverse way I was looking forward to it. I wanted to get Kathleen's scan on my co-workers, particularly Brian. "Well," I say, "we can do it on Super Bowl Sunday."

Today we lay out all the floor joists. These are the TJI composite members. Apparently they are tremendously strong and are also called "Silent Floor" because they don't squeak like the old, regular two-by-twelve joist material. "They sure burn like crazy," Vern says. When a house is on fire, he explains, these TJI beams are the first to go. All the floors fall through. A fireman's nightmare.

Vern seems tense and discouraged. Nothing fits. "Everything's too damn high!" he says, in alarm, as if he has just discovered a gigantic fuck-up, and I think back to my question before we poured the stem walls. Why didn't he shoot the elevations one last time? Anyway, he didn't—and so now we have to go and chisel sometimes as much as a half-inch channel off the pony walls where the TJI beams sit. Bud and I do this while Vern and Brian start running the stock behind us. Once the process gets underway, it goes pretty quickly.

The day warms up, the sun is out, and Brian takes his hat, coat, and wool gloves off and sets them up to look like a scarecrow. He's an irrepressible clown, but he works hard, so Vern seems content to let him goof around from time to time. In midafternoon, Bud and I are standing near the garage and Brian is working alone in the opposite corner of the house, near what will be the master bedroom closet. He is nailing off the joist hangers and singing songs from the *Jungle Book* to himself. Bud looks at me, then at Brian, then back at me. "He still hasn't quite made it out of childhood," he says. "Well, he will real quick," I say, "when his wife pops out that first baby." Jennifer, Brian's wife, is expecting their first child in mid-June.

The weather is nice, the golf course is crowded, and now it seems every fifteen or twenty minutes a golf ball lands in or near the site. One lands on the Chin Seniors' roof and shatters yet another ceramic tile. Somehow it got through the protective fence. "Incoming!" Vern calls out. He was in Vietnam for two hitches. A Seabee, he built bridges and air strips, was in mortar range a good deal, and was in Hue during the Tet offensive. He hasn't asked me what I did in the war, but he must sense somehow that I wasn't out there packing an M-16 through the jungles. It's strange to be in a golf-ball target zone. Brian and Bud and Vern all seem slightly angry with the golfers and will not help them find their balls. Brian even picked one up and put it in his nail pouch and when the guy came by, Brian said he hadn't seen it, but that he thought it landed somewhere near the outer foundation wall. I half-jokingly plead with Brian and Bud to throw back the golf balls, at least out into the rough where the golfers will find them and consider themselves lucky.

I was born into a golfing family the way these guys were born into hunting families. To my Dad, golf was almost a religion. He very nearly came to blows once with Paul Weiss, the famous Yale philosopher, who was at my grandparents' house for Sunday lunch. Paul Weiss suggested that Ike spent too much time on the golf course, and my Dad took this as a personal insult. While Vern was out in the woods as a kid with his Dad stalking elk, I was out in the woods with my Dad stalking golf balls.

One golfer comes in along the walls, but won't get down onto the plastic that now covers the crawl-space gravel because he's wearing cleats and doesn't want to puncture the vapor barrier. Reluctantly, Bud goes to get his ball for him. Bud is preoccupied anyway. On watch at the hospital, all night every night. Now Carol Anne is starting to lose her hair. I can't imagine what they are going through.

Vern has finally signed the contract to build the entire house. "I got her to put her name on the piece of paper," he says. The house is due to be finished May 31, so that gives us some time.

Beverly Chin, the client, came by earlier, and we talked briefly. She is in her late fifties, I would say, an acupuncturist and painter. She tells me she received her medical training in Beijing in the sixties. "You were there during the cultural revolution," I say. "Yes," she says. "Sometime when you have a minute we can sit down and chat about that." "I would love to go to China," I say. "Well, you can go," she says, "but not until you finish my house." I like her.

December 29

We finish up the joists in the morning. No Bud today. "It's colder than snot," Brian says. I work all morning in bare hands because Vern works in bare hands, but by early afternoon my thumbs are numb. We eat lunch out on a piece of plywood slung between two sawhorses, but it's so cold I want to get back to work just to warm up.

In the afternoon, while Vern fusses with layout, Brian and I move a huge pile of gravel from one side of the house to the other. The shoveling warms us up. We have built a ramp that goes up to what is now floor level, then across the joists on two pieces of three-by-four ply (the old formboards), and then down into the pit that will be the patio. Brian drives the wheelbarrow because he is so strong and can take such huge amounts at one time. He doesn't lose one load, although there are a couple of near misses, one almost onto the head of the plumber's son, who is on his back fitting out the drain line. The plumbers are rough-plumbing the house before we lay on the plywood—tongue-and-groove, inch and a quarter, if it's like the other house. One four-by-eight sheet weighs a ton—another good job for Brian. Too cold for much golf, so we don't get a lot of action with flying golf balls, although one golfer comes by and wants to know if Vern needs another carpenter. "Nope," he says, "we got plenty right now."

Bud doesn't show up, and Vern comments, "Must be a helluva lot warmer in that hospital." Carol Anne is still at St. Joseph's.

In the late afternoon, the plumber's son falls through one of the TJI beams that was not completely nailed off. He's not hurt but is red-faced with embarrassment as his Dad chews him out. The accident happens while Brian is elaborately recounting to Vern the plot of Wes Craven's *The People Under the Stairs*. Vern looks like he is being buffeted by an ill wind. He finally terminates Brian's monologue by sending him to rip down some pressure-treated blocks off the two-by-eights.

Now, with the plumbers here, the other subs coming by, Vern with the contract in his pocket, I am starting to have a real sense of the house. I love the look of the even rows of floor joists—level, straight, ready for plywood. The day goes by fairly fast and I half-think we will get tomorrow off, but when I ask Vern at the end of the day, he says, "No, come on by, we'll find something to do."

I drive home, have a cup of coffee and a Power Bar, and go off to play a hockey game against the high-school hockey team. Our team is made up of a selection from the over-forty group. Youth against wisdom (or so we tell ourselves). I score two goals and we win, but my legs feel like spaghetti, particularly in the third period. Hockey has always been such a passion for me, ever since I was a kid. In Boston in November, I'd pray for that string of cold nights that would finally freeze the ponds. I played all through prep school, was captain of my team my senior year, and played briefly on the Harvard freshman team before I decided that cigarettes, Scotch, and Jean-Paul Sartre were more alluring than slap shots, garbage goals, and three-man breakaways. When I moved to Oregon in 1991 and discovered there was a rink here and a city league, I bought myself the gear and started up again after a thirty-year hiatus. I thought I would never play hockey again in my life, and now I play twice a week.

December 30

The last day of work this year. A Friday. The plumbers are still on site, laying drain line and sweating copper. John and John, another father and son combo. So many years on construction sites and so many phrases and bits of lore that have stuck to me. All day I'm wandering around saying something I heard somewhere years ago: "It takes me a long time to dig a trench but I can sure lay pipe." When I say this to Brian, he laughs, so I take it he hears, as I do, the double meaning. So much of construction, particularly plumbing, is sexualized. Anything closer than one-sixteenth of an inch is inevitably a cunt hair and the smallest of cunt hairs is always a red cunt hair. Everyone "likes it tight." Plumbers have male and female fittings and nipples and reamers. Most of the plumbers I've known seem to be named John. Actually, plumbers tend to be the most twisted people in the trades, just as the roofers are the wildest, the drunkenest, and have the highest number of tattoos per square inch of exposed skin. Electricians are the cleanest, although, oddly, the drivers of the concrete trucks tend to be meticulous about their clothes and boots.

When I was working in Southern California, there was a plumber who came on the job every few months, always with the same story about how Angie Dickenson had called him up for a leaky valve behind her toilet. "Then I was down on the floor replacing the angle stop," he would tell us, as if we had never heard the story before, "and she comes in and she's wearing this kimono-like thing and she puts one foot up on the bathtub like this and I look up and she's not wearing any panties." I think he was a post-traumatic stress victim from the Korean conflict. Another John.

On this job, there is almost no sexual banter. Vern seems like he must be a strict Christian, possibly "born again." He attends a small church that happens to be my polling place, so I go in there once or twice a year or so to vote. Heavy Christian vibe. Plus, when we were trying to schedule the office party, he had two nights a week that were out of the question: Monday was Fire Department and Wednesday was Bible study. He rarely swears and if he does it's under his breath. I think I've heard him say "son of a bitch" a few times. Even today, at one point when the clean-out for the gutter drain line doesn't come out right (we are doing some of this easy plumbing ourselves), he starts kicking the thing and mutters a few old-fashioned cuss words under his breath. He does have a temper, which he takes out on either the materials or the tools. He has never lost his temper with either me or Brian. Maybe once or twice I've heard him get a little peevish with Bud, but Bud is family.

Today remains cold but with no rain. Vern comes out of his truck at 8:00 this morning wearing gloves—a first. The temperature starts out in the low twenties. I kid him about the gloves, tell him I've lost my faith. But I have my own gloves on almost till noon.

We spend six hours running the drain lines for the gutters around the exterior perimeter of the foundation. Bud and I do that while Brian digs trenches. I think Brian's love affair with the shovel is wearing off. I tell him they call shovels "Mexican banjos" in L.A., and then regret it immediately, as it plays into all his racism and I am the initiator. I have to learn to keep my mouth shut. Particularly with Beverly Chin as the client—there's such a temptation to go into Asian-American dialect. Brian and I did that for a while when we were skip-sheeting the roof on the Baxter job and I regretted it then. These things keep popping out of my own racist mouth. On most construction sites it's nigger this, beaner that, pussy this, fag that—on and on. This site is more sober and serious than any I ever worked on. No one has told an off-color joke, for example. No one talks about how much they drank or gambled over the weekend. At break today, because of the seating arrangements, I was a bit apart from the rest of the crew. Brian, Bud, Vern, and John and John, the two plumbers, spent twenty minutes swapping hunting stories and I sat off, near where the front door will be, smiling as if I understood what they were talking about, and eating my yogurt.

We knock off at 2:30 because there is nothing more to do and now, after the New Year's break, we may have a few more days off while the electricians and the heating people and the insulators do their thing. Then it's balls to the wall all the way through the framing. I haven't worked back-to-back five-day weeks in months, it seems like. There may be a few ahead in January.

January 11

I T'S BEEN almost a week and a half since we last worked. I spent the time off staring out at the freezing rain and wondering how anyone could ever work in such crap. Talked to Vern a few times during the time off and he was frustrated about the hold-ups. "I just can't seem to get over this hump," he told me at one point. Inspectors, subcontractors, some changes Beverly came up with in the way the plumbing was laid out. I hardly picked up a tool during all this time. The closest I came was going to Fred Meyer to spend my $50.00 Christmas gift certificate. I got a new framing hammer, a wooden-handled Vaughn "California Professional" with a large waffle head and a straight claw. It's a 23-ounce instead of the 24-ounce I've been using. The handle is slightly longer too, and less round. Also, I got a new combination square, a new wrist support, new glove liners, and a giant clamp. The total came to $50.45. Pretty close.

Today, Vern finally calls about noon and says we are ready to start laying down the floor sheeting. I get to the site around 12:45. Pouring rain. I open the back of the car, slip into my rain pants, belt with the new tools, rain jacket, and Gilligan hat. I've also broken out my winter rain boots: cheap, Kmart, rubber lace-up boots, but with some kind of synthetic lamb's-wool liner that feels like heaven. At least my feet are warm and dry.

The others are already at work. Everyone in yellow raingear. Brian and Bud are working jackhammers, enlarging the crawl holes to eighteen inches—one of the changes the inspector demanded. Vern is starting to lay down the sheets of plywood, so I get into stride by helping him. Pretty soon Brian comes over and works with me, while Vern moves on ahead. Bud, once he finishes his enlargement job, sets up the compressor and starts nailing off the sheeting with the nail-gun.

Heavy rain comes down all afternoon. Vern remarks at one point that the last three days, as he sat around waiting for inspectors who sometimes never even showed up, the weather had been perfect—dry, warm, and sunny. "It never fails," he says, taking this rain as yet another proof that the once again the gods of of construction have stacked the odds against him. Now the forecast, which I see tonight when I get home, is for rain, rain, and more rain. It's funny, when you're inside looking out at weather like this, you cannot imagine how anyone could work in it. But once you're out in it, moving along, it begins to take on a reality all it's own.

Brian works without rain pants and, at least for part of the afternoon, with neither a rain jacket or a hat. A real Oregon kid. We don't have much time for small talk. In fact, he's rather reticent today. About the only thing he tells me is that he took a swim in the King River when his drift boat capsized while he was fishing with his friend on the day after New Year's. They lost some of their gear. "I bet it was cold," I say. "I didn't even notice," he says, "the adrenaline was pumping so hard."

Around 3:30, Vern asks about the time (he's never done that before), and then says it feels like 6:15. He calls everything to a halt a little early when the rain is just pouring down in buckets. "More of the same tomorrow?" I ask as we pack the last of the tools away into his truck. "Yep," he says, "eight o'clock. We got to really start moving on this thing. Too damn many delays." I wonder if he goes into a penalty phase if we go past the thirty-first of May without completion.

As I walk back to my car I notice he has a construction sign up: MARSHALL CUSTOM HOME CONSTRUCTION—THREE GENERATIONS OF FINE HOMEBUILDING. Makes me feel right proud. It feels good to finally be back to work. For the first half hour or so this morning, it's like I've forgotten everything I ever knew about building. I'm all thumbs, slow, awkward, dropping numbers, missing cuts, slipping on the wet TJI joists, but then slowly, as if a fog is clearing from my brain, it all comes back to me, and by the end of the day, it feels in some ways like I have never left. Tomorrow we'll do it again.

January 12

The first full day back. Forecast is for rain, and in the morning a light drizzle is falling. No golfers. We pick up right where we left off yesterday. About a third of the way across the house, moving north to south, laying down the tongue-and-groove three-quarter-inch plywood subfloor. Brian and I cut and place the sheets down on the glued floor joists, Bud nails them off with the nail-gun, and Vern follows behind, beginning to lay out the walls. The work goes smoothly. Brian is happy and rather quiet; in fact, he's been quieter than usual the last two days, maybe because he's absorbed in what he's doing.

We are all in raingear, but after half an hour the rain stops, and then it holds off for the rest of the day. I work the whole day with my rain pants on, with jeans and long underwear underneath, and my pants are soaked with sweat by 2:00. Still, working with dry materials feels almost like a vacation. The chalk lines hold, the glue squeezes out easily, nothing is slippery. The only mishap occurs toward the end of the day when we are wiggling a large sheet of plywood down over a line of pipe stubouts. The sheet gets hung up on the thick edge of a drain pipe in the far corner of the master bathroom, and when it finally slips free, my fingers are underneath it, and the whole sheet comes down on my hand.

Brian thinks this is the funniest thing he has seen all day. I tell him the funniest thing I have seen all day is when he went to use the Skilsaw and failed to notice that it was unplugged and all he got was a couple of dry clicks on the trigger. "Spark don't jump that far," I said, which is what I always say. Some wiseass said it to me twenty years ago. The unplugged Skilsaw is a recurring event in construction. Drills and Sawzalls come out and get plugged into the one extension cord in the area, and sooner or later someone will pick up the Skilsaw and click the trigger and . . . nothing. If you're working hard and fast and things are not quite going right, this can touch off a burst of curses at the universe in general and the construction gods in particular.

At the end of the day, we start framing up the first walls. The two-by-sixes are soaking wet and heavy—definitely "pond dried."

January 13

Friday the thirteenth and it rains like hell. I make a point of leaving the house on time and am the first one to the site. When Vern shows up, I'm in my raingear with my belt on and ready to go. I have a cold. Sore throat, achy joints, feverish. Kathleen tried to talk me into staying home today, and when I said absolutely no way, she told me I was as stubborn as an old mule. I always work when I'm sick like this. As long as I can still walk. It doesn't seem to make the illness any worse and it changes the tempo and feeling of the work. When I used to work with hangovers time always seemed to go by a little faster. Every step is a bit of an effort.

By 8:30 we are in full swing in the drenching rain. A real "frog strangler" as I heard some old-time meteorologist refer to it on the radio on my drive over. The harder it rains, the more cheerful and chatty Brian becomes. Still, we work together well. Vern does the calculations and gives Bud the measurements, Bud cuts the heavy-as-lead two-by-sixes on the old, blue radial arm saw, a relic from the fifties that was just set up on the job yesterday. The saw has been in the Marshall construction family since Vern's grandfather's days. Vern has built a little roof over it, so Bud is out of the rain while he cuts, but I think Brian and I are warmer because we are moving around.

We are working on four long sections of wall—studs, window and door openings, headers, trimmers, and short blocks—virtually all the framing of the walls except the second top plates will be assembled and nailed off while they are still on the floor. And the wall sections will be sheeted and Tyveked before they are raised into position. They will weigh hundreds of pounds each, who knows, thousands maybe, and will be raised by wall jacks. I have never framed walls of this dimension and raised them with jacks, so this will be a new one for me. Next week sometime, I guess, we'll start raising them up.

The rain continues to pour down in a steady, soaking sheet. A few golfers shuffle by with umbrellas and full foul-weather gear. Mostly old guys. "You guys are crazier than us," we shout. Water splashes up into our faces as the heavy studs are dropped down into deep puddles that have pooled up onto the plywood subfloor.

We eat lunch crammed into the cab of Vern's truck, the heater blasting, all in our wet, rubber raingear, the windows misting up, the talk straying to old flood tales, houses Vern remembers from his childhood, bobbing along down the river.

Finally, at about 2:00, with the walls framed, Vern decides to cancel the rest of the day. This is unprecedented. The rain, if anything, has gotten heavier. We're going to be sheeting the walls with a kind of particle plyboard that swells up badly when wet. One-half-inch wood, it bulges to three-quarters of an inch around the perimeters of the windows and doors, making the trimwork impossible to fit.

On the drive home, I realize my clothes are totally soaked, pants, shirts, underpants, sweater, shoes, socks—everything. My belt weighs twice what it normally does. I get home and hang it all up to dry, as best I can, suspended from chair backs over electric heaters around the house. By 3:00 I have taken a hot shower and I climb into bed and doze through the rest of the afternoon, in a cloudy paradise of soft, warm comfort.

In the evening, Kathleen and I go out to dinner at Vern's house. I'm still sick. This is the long-awaited holiday party that was postponed several times due to scheduling conflicts and Carol Anne's illness. Now we drive up through a big ranch-style entrance, way out on a rural road in the Riley Valley. Elk horns and the name MARSHALL adorn the top of the arch above the entrance. The house is set back from the road several hundred yards on a wooded hillside. Lights are glowing from the large front windows. We are the last to arrive. Vern is there, his wife Connie, Bud and his wife Debbie, and Brian and his wife Jennifer. Carol Anne has stayed with her other grandparents, partly because I told Bud today about my cold, and they are anxious to keep her away from any free-floating germs.

We go right to the table. The big front room, with dining area and sunken living room, is done in dark wood with a bright, almost neon-blue carpet. Knick-knacks, most of them in shades of blue, adorn the walls. The meal is brought out by the women: lasagna, green beans, salad, bread sticks, and apple sauce. No alcohol is served. I always pictured a whiskey ritual at this party, and an awkward moment when I would have to turn down the offered glass, and would finally be exposed as the dreaded alcoholic in their midst. But no, in reality Connie tells us, "We have water, milk, or pop." Once all the drinks have been poured into the heavy, green glass goblets and the chatter dies away, Vern intones a long, solemn prayer, the words of which he seems to be creating as he goes along. "Dear Lord, we thank you for the many blessings of this year, for the food and shelter which we have enjoyed, for the work and for the friendship, and Lord, we ask you now for your continued guidance and blessings in the year ahead. Hear our prayers Lord, for the health of little Carol Anne and also for this new baby, for the work and business to come, and for your ongoing help and protection. Amen."

The reference to "this new baby" is for Brian and Jennifer, who are expecting their child in June. Just today, they went in for the sonogram of the fetus. Brian tells us when we first arrive that he has been able to see the little infant's mouth on the video and that it was moving! "Another Brian!" Vern says. "Already talking in the womb! Jennifer—you're in for big trouble." We all laugh, maybe a little too much, but Brian, in his irrepressible good nature, laughs with us.

The conversation drifts through safe subjects—no current topics, no debates, no politics. Brian tells the story of his recent dunking in the King River and the amazing recovery of almost everything in the boat as well as the boat itself. We talk about families, the weather, techniques of discipline. Brian tells about the time his mother broke down the door of his room and cut the power cord to his TV with a bread knife. He was fourteen at the time. Vern says that the best thing he ever heard of was a Native American tribe that employed a single individual called the "whip man" to take care of every family's disciplinary needs. Tales of disastrous boy-scout camping trips come up. A mood of good will and fellowship develops around the table. Kathleen keeps the ball bouncing in her skillful, family therapist way. I am still feeling sick and congested and struggle ahead, drinking glass after glass of water.

After dinner, Vern brings out a big album and shows me pictures of the houses he has built, houses going all the way back to a flat, one-story ranch house from the mid-fifties that, he says, was the first house he ever worked on. I wish I had a similar album of my own. If I go back to San Francisco and L.A., I should spend some time driving around, taking pictures of all the houses I can find, the ones I remember. How many houses have I worked on in my construction career: twenty? thirty? Somewhere in there.

As the clock ticks away toward 9:30, we sit around in a circle, in giant, stuffed chairs and couches. The family dog, Yogi, entertains us with his yowling songs. The men tell hunting stories and the women tell about the time the big fire department tanker rolled over with both Vern and Bud inside. I heard this one once before from Bud when we were waiting for the concrete.

When we drive home, in the still-pouring rain, Kathleen wonders if these people aren't the real normal, well-adjusted people we have heard about but never met. So different from the hip, slick, and cool, coastal, collegiate yuppie/arty people we know in our own circles—but genuine too, and appealing and warm and full of laughter. During the three hours we are in their home, there is not one mention of anything controversial. It is almost like an extended story hour. "Isn't that maybe a Christian thing?" I say to Kathleen. "I mean, maybe we stereotype these Christians too much. Maybe we have this sense that all Christians are right-wing, fundamentalist, right-to-life blockheads, when in fact many of them are just good people, good, old-fashioned American people with those same home-grown values we see on old Gary Cooper movies."

There is something of the Gary Cooper about Vern. Tonight he shows nothing of the deadly serious, no-bullshit construction boss. He is tall, thin, lanky, attractive, charming, witty, quick to laugh. Kathleen is very taken with him. Considering how unpremeditated and chancy my employment has been—I needed a job and he needed a carpenter and so I ended up working for him—tonight it almost feels like a blessing to be working with this crew. Now two days off to recover from my cold.

January 16

MONDAY. Full Moon. My cold has shifted now from the early stages of ache, sore throat, and chills to a plain stuffy nose. I drop an Advil before getting to work and this seems to ease the symptoms. The day is cloudy and cool and uncertain but with no real rain. A few sprinkles in the A.M. The sky shifts and changes through the day with periods of sunshine that feel almost blinding. What a change. I'm wearing just my jeans and lightweight long underwear, a tee shirt, a turtleneck and jacket, my sponge hat, and no gloves. The Kmart winter boots are still working great. I added removable insoles to them this morning and they seem to fit better. Brian bought a pair of enormous, yellow rubber boots that slip over his size fourteen shoes. They look like two miniature drift boats barging out at the toes from under his yellow rain pants, which he wears all day, even though there is no rain.

We are sheeting and then Tyveking the walls, which lie flat on the floor ready to be raised once they are nailed off. The process seems quick and easy, especially in the dry weather. We cut, place, and nail the particleboard, then spread out and staple the Tyvek, a white, lightweight Dupont product that acts as a vapor barrier and replaces the old fifteen-pound felt "tar paper" we always used to use. The product logo, a house-headed cartoon character wearing a Tyvek jacket, has become "Tommy Tyvek" or "The Gable Head" in our ongoing banter. The only other cartoon character we can think of in construction is the Pink Panther, who makes occasional appearances on rolls of pink insulation.

A few minutes after lunch, Connie and Debbie appear with Carol Anne. I have never seen the little girl before. She was at St. Joseph's this morning for an MRI and is now heading up to Portland with her Mom and Grandmother for the second of her chemotherapy treatments. A diminutive and irresistible fourteen-month-old, she is still groggy from the anesthetic they gave her for the procedure this morning. She has on a big, raspberry-colored, floppy hat to cover up her head, which has been going bald from the chemo. Debbie holds her and then hands her to Vern, and she snuggles into his arms, managing a shy smile in my direction before burying her face in his shoulder. We stand around and "ooo" and "ahhh" for a while. Debbie takes her back and shows us the "port" they have set into her main artery so that she can be instantly hooked up to the drugs. Then she pulls the back of the hat up to show us a long, jagged, pink surgical scar that reaches from the base of her neck all the way up to top of her tiny, pale, bald head.

I feel the tears start to flood into my eyes and turn back to work, memories of Shannon as an infant mingling with images of my parents' deaths. I wonder through the afternoon if this family is going to have to face the death of this little girl. It all seems brutal beyond words, and they seem, in contrast to the brutality, the most courageous of people. Debbie, the poor mother, has a rough, abrasive edge and I think, well of course, and why shouldn't she? Later, Bud, the father, checks in. He will stay down here and work while they are up in Portland for the treatments. Through the rest of the afternoon he seems quiet and contained, though he makes occasional remarks about certain insignificant irregularities in the work. I remind myself not to take it personally. This guy's got other things on his mind.

Halfway through the afternoon, Bud and Vern raise the biggest of the walls with "wall jacks"—two large, red, racheting devices that slip over long two-by-fours. The operators crank on bent, yellow handles and this huge, heavy, forty-foot long, nine-foot high, sheeted and Tyveked, pond-dried, two-by-six wall raises up to plumb within a minute or two. Every other wall I ever built was raised by brute strength. This process seems so remarkable and efficient. Plus we have turned a corner in the progress of the house—the first wall is up! It is the south bedroom wall, and when Beverly comes out to look at it later, she has immediate concerns about the lack of light. Perhaps we will have to cut in more windows. "Get out your change-order pad," I tell Vern.

Mostly Brian and I work together—fluid, easy, happy in the task at hand. At the end of the day I'm tired, but relaxed, and after a shower and a cup of coffee, I go to my evening meditation class and attempt to meditate in a half-somnolent state, measuring, cutting, and nailing down sheets of particleboard in my always-busy mind.

This is just a small class, eight or ten people who get together at the church every Monday evening to learn meditation. I heard about the group through an AA friend, and it has become a regular part of my life. I've always wanted to learn how to meditate, and now I'm discovering how difficult it is. The principle is to keep your mind open, focus on repeating your mantra, and let the thoughts drift by. What you don't want to do is get caught up in any one thought, an almost impossible challenge for me anytime—but particularly after work. I'm saying my mantra, but mostly I'm nailing off sheeting, I'm hanging Tyvek, and I'm listening to Brian retell the plot of *The Fugitive*.

January 17

A wicked-cold, wet day. When we get to the site at 8:00, there is slush on the wood, and some of the stray scraps of two-by-six and particleboard are frozen to the plywood floor. I have on everything I own, and feel overpadded and slow. My belt feels like it's full of stones. We pick up right where we left off, but now we are working in the rain, which comes down slow and steady, just on the high side of freezing.

The golfers emerge out of the mist like batty old men in their rain outfits, just like us, but hacking those little balls around the soggy green turf. It's as though they are paralleling us, but in another universe. I know none of them by sight, but I have the feeling that it is the same old guys every day, regular foursomes that meet three times a week in the early morning. My Dad would have been one of them on a clear day. He didn't much care for playing in cold, wet weather. With his heart condition and poor circulation, he had to quit when the temperature dropped below fifty. I remember once, toward the end of his life, he had to pack it in on the fifteenth hole at Dedham one raw November afternoon, and drive the golf cart back to the clubhouse with his purple, frozen hands stuffed into his knitted clubhead mittens. My own purple, frozen hands I keep today inside thin, polypropylene glove liners. But that only works for a while. Once the gloves are wet, they do no good, and they're wet soon enough.

At 10:00, I go off to the dentist for work on a tooth that broke off while I was eating a Power Bar on my way to a hockey game a few weeks before Christmas. I am happy to sit for an hour in a warm dental chair and listen to the elevator music and have Dr. Giles build up the stump of my broken tooth. What's a little oral pain if you're warm and dry? I'll go back for the crown in ten days.

After lunch it's more of the same. No break at all in the weather, though it must be a few degrees warmer than first thing this morning. Brian and I frame walls, sheet them, lift them, nail them off, move on. At some point, I'm not sure exactly where or when, I tweak my back. It's a muscle in the lower-left quadrant, and by the end of the day I'm having trouble bending over. The last thing we build is a huge header made of four fourteen-foot pond-dried two-by-sixes. The thing is massive and cumbersome and has to go up into a spot on the front wall about ten feet off the floor. It's the header for a long row of transom windows. I have my doubts about lifting it. Because we're at the end of the day, Vern says leave it for the morning. Thank God.

I drive home obsessing about my back. I've thrown it out before but not for years. Some of the walls we lifted today were heavy, heavy, heavy, and we only used the house jacks once. I give my back a hot shower when I get home, then sit on the couch in front of the TV with bags of frozen peas pressed into the sore area. I've done all I can do—now I just have to hope for the best.

January 18

Wednesday, and it feels like the deep darkness at the center of the week. I wake up stiff and barely able to get up off our bed, which is on a low platform six inches above the floor. I drink coffee, read the newspaper, and then spend some time messing with my nail bags, shortening the suspenders and widening the belt. For years, I wore my nail bags slung from my hips, which is the way they are designed, but then some chiropractor said that carrying that heavy belt pressed down into those nerves just below your waist could turn you into a total cripple. So I started wearing big, wide, red suspenders to hold the bags up, and loosened the belt so that they would swing free around my hips, and so that my shoulders would carry most of the weight. This took a bit of getting used to, but after the first few weeks it came to seem completely natural. These days, wearing all these clothes, the belt has begun to sit tighter around my hips, hence the

[54]

adjustment. It was probably overdue. But my lower-back pain has made it imperative.

I toyed with the idea of calling up Vern and telling him I couldn't work, but there is a stubborn part of me that refuses to back away from work unless it is completely impossible. So far, working for Vern, I have not lost one day to sickness or injury. I want to keep it that way. I tell Kathleen, as I am about to go out the door, "This is the day that will really separate the men from the boys." Her response: "I don't want to hear any of your macho bullshit." All morning she's been telling me to call in sick and take a day to let my back rest. She always pulls my covers on this one. I go off half-worried I won't be able to move around much at all. But I take a couple of Advil just as I drive up, and once I am out there, up and down ladders, hauling the heavy lumber around, my body seems to loosen up.

Brian is in one of his obnoxious moods at the beginning of the day, making cracks about my work, watching to see if I screw up, telling me how to do things, bossing me around. He's starting to drive me a little bit crazy as we frame up the interior walls for all the bathrooms and closets adjacent to the master bedroom, and I find myself singing that old song under my breath: "Oh Lord, ain't it hard to be humble, when you're perfect in every way . . ."

It's another mean, cold day, but not raining, at least for the first four hours. In the afternoon, the rain sets in hard and the floor is mostly awash in an inch of water. I'm nailing up two-by-four "California corners" (stud-block-stud) for all the wall junctions and do it on my knees in deep puddles of water. The water runs up the legs of my rain pants and I am soon soaked. But warm, as long as I keep working. Thank God for the Shetland sweater. At one point, when we get caught up with Vern, who is laying down the plates and doing the layout ahead of us, we energetically clean up the whole site, just to stay warm.

Bud seems to be in a foul temper today and I give him as much slack as he needs. I cannot imagine what torment it must be for him, day to day, to have a child that sick. Vern seems gloomy and off in his own world, swathed in yellow rain garb, his old dilapidated nail bags hanging from his waist (no suspenders for him—for forty years he's been walking around with those damn things strapped into his skinny hips), scratching his head, laying out plates, cursing under his breath as he discovers pipes that have missed walls and incomprehensible numbers on the plans.

I work all day with my bare hands wet and hardly notice it. I must be almost desensitized to the discomfort. My back is stiff and hurts when I get into certain positions, but at least it is no worse. Still, when we get to 4:10 and start rolling up the tools ten minutes early, I am elated. Just the feel of getting into the car with the heater on is enough pleasure in itself.

Emily has invited us to dinner. A cup of coffee, a few minutes by the fire at Emily's, a dinner of hamburger and mashed potatoes and peas and apple pie with Shannon, the warm glow of simple family life—it could be boring, but in contrast to eight hours shuffling around with a bad back on a wet, cold construction site, it's damn-near paradise. At the end of every day, there's a little mini-weekend where you can just sit around and do nothing at all. I live for that.

January 19

A foggy, cold day, all day, the temperature hovering around the high forties but with a wind that works right through you if you stand around too long. I try to keep moving as much as possible and have on my heaviest long underwear. No rain. I mutter prayers of thanks under my breath.

We are framing the walls in the southeast quadrant of the house, the bedroom end with a maze of closets, bathrooms, and utility rooms. The work is straight ahead and simple, cutting plates, laying up the studs. Most of the work is "stick framing," which means that the walls are built in place rather than built on the ground and tilted up. This is partly because there is not enough room anymore at this end of the house to lay down a nine-foot wall, and partly because there are so many pipes sticking up that it would be hard to drop a fully framed wall down over them.

The last step in this wall-framing process is to plumb up all the adjoining walls in an area, brace them off, and then nail on the second, double, top plates, which overlap the first top plates and tie all the walls together. When we get to the point where we are ready to plumb the walls, I check everything and find that the entire house, exterior and interior walls alike, is off by about a quarter inch over six feet, leaning slightly to the north. I tell Vern this, and he comes to check for himself. We use the rattiest old six-foot level, but it is apparently accurate as it reads the same both ways (the test). It must have been on the Marshall construction company truck since the days of Ike and Mamie.

Vern walks around, checking, looking, slapping, then slamming the old level up against the studs at various locations; he fumes, he storms around, he mutters under his breath, he's furious, but at what?—the gods themselves, really, that they should have created a world where such a thing is possible. "Every damn wall is out," he snaps. "How can that be? You make 'em all square and level and you set them up and every damn one is out!" He gets himself madder and madder, he's working himself into a fit, slamming the level against one stud after another, as if, it almost seems, since he can't move the walls he might at least knock some sense into the stupid level. Brian has said before that Vern, when he's mad, reminds him of Donald Duck, and it's true. You half expect him to just explode into a burst of quacking expletives. But he's never mad at any one person. He's mad at the tools, the materials, the traps and uncertainties and irregularities of the process itself.

Later in the day, when Bud cuts a whole sequence of two-by-six studs for a sloping wall at an eighteen-degree angle instead of a thirty-degree angle, Vern is not so much mad at him, as disappointed, even embarrassed. But it's true, we *all* make mistakes. I say sometimes as a recurring joke, "Yeah, even I made a mistake myself once, but it was way back in the seventies sometime." The truth is, I make a few mistakes almost every day. Today I cut several top plates wrong, I dropped some nails, I set a board up backwards. It seems inevitable, the builder's torment—human imperfection. But when Vern finds these screw-ups, particularly his own, it seems to drive him into an almost apoplectic frustration, as if the organization of the universe itself was somehow faulty. Brian, at the other end of the line, makes mistakes continuously and tries to cover them up as best he can, to hide from them or make excuses for them.

And Bud, who, like all of us, makes his fair share of blunders (the wrong angle on the studs today one of the worst I've seen) has an irritating habit of scrutinizing my work or Brian's and then trying to correct it, if he is standing around with nothing to do. Today, for example, I found a wall-block arrangement that was improperly nailed and out of square. I set out to correct it and suddenly I could feel his shape right beside me, watching me. I took out a sixteenpenny nail to square the stud, and drove it in, toenailed, as we say, that is, driven into the base of the board at an angle so it will catch the plate below. And Bud said to me, "You don't really have to use a sixteen, it's too big a nail. Always use an eight for that kind of thing." Sixteens are about three inches long, eights are about two and a quarter inches long. I felt a mild surge of anger and could imagine myself saying what Brian sometimes says under his breath and out of earshot at similar moments of correction from Bud. "Hey, when you're the one that's signing my paycheck, then I'll be glad to listen to whatever the fuck you have to say." But I don't say that.

[58]

I look up and smile and say, "Oh really? Okay. Well, thanks for the tip."

Maybe it's hypocritical, maybe it's even shaded with a bit of sarcasm, but still, my restraint feels almost spiritual. And also I think of Bud's baby, who is due home today from Portland after her chemotherapy session up there, and I think—hey, this guy's got a major load on his mind, let him blow it off whatever way he has to. If it means coming around my work and nit-picking, well so be it and God bless him and he's welcome to nit-pick on me any time.

The days go by. It's already Thursday. Tomorrow's Friday. This will be the first five-day week in a long time. Miraculously, my back seems to be getting better, although Vern and I joke a little about being old men and hobbling around at home every night. Sometimes when I get up out of a chair, it's hard for me to stand up straight.

January 20

Another cold, foggy day, though the sun does manage to crack through the clouds from time to time. No rain. It's Friday and the hours seem to drift by at a snail's pace. I have tried sometimes to work with no watch, to lose track of time and hence lose that clutching, anxious feeling of wanting the day to be over. But it doesn't work for me and I always go back to wearing a watch. Brian does not use a watch. Vern has the watch for the job and calls out the two formal breaks we have, coffee-break at 10:00, lunch at 12:00, every day exactly the same. And I spend the day either looking at my watch or not, depending on my mood and level of impatience.

Today I check it first around 9:30 and know we have another half hour till our first break. When 10:00 comes, it is always Vern who calls it out. "Coffee time," he says in a voice that sounds both weary and relieved. He always has coffee out of his solid-metal, green Stanley thermos, and a muffin of some kind. Bud has a Pepsi and a candy bar. He's into sugar. I usually have coffee and a yogurt. Today, Brian has no food for break and just sits on the stack of two-by-sixes and talks. Later in the morning, he gets grouchy and touchy and difficult and I wonder if he's had a fight with his wife, and then I think it's because he hasn't eaten anything. For lunch he has brought his father's giant propane barbecue in the back of his truck. I help him carry the thing over to the site. It has a five-gallon propane tank. He slaps on six burger patties and runs to the store, leaving me in charge. He asks me to check the burgers in four or five minutes. "You got a flipper?" I ask. He makes a motion with his fingers—apparently I'm supposed to do it barehanded. So I end up flipping these burgers with my fingers, feeling mildly resentful that I'm spending my lunch-break cooking his lunch while he's off buying his buns.

After lunch, the day really starts to drag. The early afternoon is our longest run without a break. Coffee is from 10:00 to 10:15 and lunch from 12:00 to 12:30. We knock off at 4:30. These times are very strict and vary only a minute or two in either direction from day to day. As for arriving at work—the time is 8:00. Vern is always at the site between 8:00 and 8:05. Brian is hard to predict but always pulls in before 8:30. Bud usually doesn't arrive until 8:30 or 9:00. I am never more than ten minutes late and hate it if I am even that late. Today I left home early because it was Friday and because I wanted to be the first one on the site, to be the one who gets the honored parking spot by the Sani-pot. And I am first. By the time Vern has arrived I am ready for work and we go into the morning tool roll-out routine.

Because it's Friday and because we finally have something of a house to straighten up, we knock off work a little early in the late afternoon and clean up the site, scrap out the wood blocks where they have accumulated under the sawhorses, sweep, pick up the stray scraps of Tyvek and chipboard. It feels good, that last half-hour of clean-up and the thought of a weekend ahead.

Quite a week. Monday morning there was not a wall standing and today by the afternoon we seem to have most of the walls of the first floor done. All except one last wall near the garage and another, very complicated wall, with six angled facets, which will enclose the big front door. More head scratching for Vern and probably a few Donald Duck tantrums. The six-foot level will suffer a few more nicks and scratches. TGIF.

January 23

I AM NOT at all happy today about being back at work. Even though I am the first one at the site and am walking around with my belt on when first Brian, then Bud, and finally Vern arrive. Vern has been picking up supplies, including nails for the nail-gun. Two boxes, which cost almost a hundred dollars. He makes a comment later about not overnailing anything. And again, at another point in the day, when he drops a nail and Brian says, "We'll be taking that out of your paycheck," he makes a point of telling us all that he doesn't get a paycheck, just what's left over at the end of the job. Am I supposed to feel sorry for him? Of course he's not doing this for fun! And it's true he does put the big risk into it, a risk I would never want for myself. Still, I would guess, on a well-run job, with no major fuck-ups, the "what's-left-over" part might run as high as $20,000 to $30,000— not a helluva lot for a man at the top of his profession to make for five or six months of his life and all that stress.

I feel like a geek today. The work is getting routine. We frame up the last of the exterior house walls and then go to the garage and stick-frame those walls. Now Brian has a back problem and is moving at about one-third pace, completely the opposite of his usual three-pieces-of-plywood-at-a-time self. Bud is surly;

Vern for some reason makes me feel guilty whenever I look in his direction. He's wearing gloves, so it must be cold. No rain, however. Time creeps by even slower than Brian with his bad back.

At one point, I come to a stud on layout that is blocked by a conduit for the TV and phone lines. This is on the south wall of the garage. I start to cut a notch out of the stud to accommodate the pipe and Vern sees me (he does seem often to be watching things even when he appears the most aloof) and stops me. Tells me to shift the layout around to accommodate a box that will eventually hold all the electrical and utility wires. A few minutes later, Bud finds the half-notched two-by-six and asks about it. Brian tells him I was going to notch it around the pipe and Bud laughs, as if that's the stupidest thing he's ever seen. Even though he was the one who put the layout mark down on the plate in the first place! And this totally pisses me off. "I don't think that's funny," I say, and I start to call him on it, and then think better of it. What's the use? We're back into that whole thing about who makes mistakes.

I think in some ways, because I am different, older, shorter, or maybe just paranoid, because I don't hunt, because they suspect me of being pro-abortion, anti-Gingrich, pro-black, pro-feminist, anti-Christian, because I hate Rush, because I'm a left-wing radical . . . I don't know, I feel like I'm becoming the scapegoat of the job, more and more, both the object and the subject of a little in-joke they have among themselves, the real carpenters, the real cowboys, the real men. Although later I realize that in the dynamic of four on this job, my closest ally will always be Brian, the other worker, because nothing will ever split up the father and the son. And I have to remind myself not to get into those three-against-one deals against Brian, because it will inevitably come back around when it's three against one against me.

Brian spends hours today talking to me about some jewelry selling and buying pyramid scheme he's involved with, pestering me to join up and at the end of the day he hands me a video tape and tells me to watch it at home. It's called *The Silent Giant*, which is exactly what Brian is not. I have no time to watch it, even if I had any desire to watch it, which I don't.

I go to my meditation class and the subject for today's discussion is "one-pointedness," which means having all your attention focused on one thing at a time as you do it. This is supposed to help when you meditate. I say during the discussion that I am the most one-pointed at work, and the least one-pointed at home. But is this true? It seems that my mind is always wandering while I'm at work, and lately it is often set into a vaguely paranoid state about my fellow workers. It certainly was today! I'm exhausted already, and it's only Monday night.

January 24

Again I'm the first one to work. Brian comes cruising in close after me and starts putting more pressure on me about this jewelry scheme of his. I tell him once and for all I'm not interested and give him back the unwatched video. He seems vaguely hurt by my lack of cooperation but I tell him, "Look, if it was a tool catalogue, maybe, or a book catalogue, or a hockey equipment scam . . . but jewelry. I mean, I'm not a jewelry kind of a guy." Hopefully he gets the point.

Vern arrives towing a trailer with sack cement, the big extension ladder, a wheelbarrow, and a hose. He has to pour a small concrete pad out in front of the front door to accommodate the posts that will hold a massive protruding beam that will, in turn, hold up the porch overhang and the whole south-central portion of the roof.

It's not the cement that bothers me, but the sight of that large extension ladder, because it means that soon we will be getting to the higher elevations. "A walk on the wild side," I always call it, which for me seems to be everything over fifteen feet. Before this job is finished I will have my moments of hanging out at the edges of the roof, adrenaline pumping, fixing the fascia boards and the frieze blocks and the barge rafters. For now we are only sheeting the walls with chipboard and no higher than twelve feet, which feels relatively tame.

Still, I'm in a foul mood this morning. Brian notices it and immediately thinks it's his fault, maybe because of the jewelry hustle. "You've been awfully grouchy lately," he says, and he's right. I'm not sure why. "I'm morbidly depressed," I tell him. And then try to explain further: "Don't take it personally," I say. "It's one of my cycles. Depression is a thing creative people have to go through." Later Brian comes back to the back wall of the garage where I am nailing off sheeting with the nail-gun, and he says, "I bet you're glad you're not a clown." I guess he means because I'm so depressed, and if I was a clown I'd have a hard time doing my job. But it's true, I guess, in a way I am the job-site clown, although Brian is a close second.

So then I tell him about the time in 1965 when I was with Star in our apartment in Cambridge and we were trying to get to sleep and this din erupted in the apartment downstairs, noise, singing, music, raucous laughter, and finally someone thumping on a piano. It was well past midnight and even though I felt young and vulnerable as a worm I knew it was my job to go down and lodge a complaint. When I got down, there I found this crazed, stoned hippie at the keyboard; he was about the first real-live hippie I had ever seen. He said to me, "You know the clown—always laughing on the outside and crying on the inside." Brian liked this story and kept referring to the punch line throughout the day.

The work went pretty well; we have the garage walls all up, framed and sheeted, we have a massive twenty-four-foot, six-by-eighteen gluelam beam running across the middle of the garage, and at the end of the day Vern is starting to lay out the "Silent Floor" joists for the second floor. Brian and I spend all day— when we are not called away on special assignments—cutting, setting, and nailing off particleboard to the exterior walls.

The day turns beautiful by mid-morning, the best work day of the year so far, with a warm sun at lunch that makes me feel lazy and stupid. The depression lifts. Maybe it's Brian and his nonsense. We have three ladders and Brian has given them all names. One, a rickety, wooden stepladder, is called "Shaky Jake," and when you're working on Shaky Jake he calls that "taking a ride on the shaky side." Then there's a fairly nice, new, two-sided, yellow, plastic-and-aluminum electrician's stepladder that, because of its color, he calls "Old Yeller." And finally, there's a rather complicated all-metal combination ladder that can be either a long or short stepladder or, when readjusted, a full extension ladder. This last one he calls "The Tin Man." Today the Tin Man had a pebble stuck in one of his extenders. Bud and I messed around for a while, fixing him, and we had a brief three-way discussion about what it was the Tin Man lacked in *The Wizard of Oz*. Consensus was that he lacked a heart. As a result of that brief discussion, Brian, off on his own nailing corner clips into the beams across what will be the garage doors, starts singing all the songs from *The Wizard of Oz*, and so in spite of the tenacious depression, I finally had to break down and chuckle to myself. Brian told me earlier in the day that he hadn't been depressed since the fifth grade.

Vern seems rather happy-go-lucky today, so the job must be going well. Toward the end of the day, he's walking the walls, standing ten feet up, balanced on a two-by-six, nailing in a top plate by just bending over. "What are you doing up there?" Brian asks. "I've come up in the world," Vern answers, and you feel like it's something he has said for years on such occasions. And it is a milestone. We're already setting in place the first skeletal outlines of the second floor. At the end of the day, I'm exhausted again.

January 25

A good day at work. Cold in the morning, with traces of ice on the wood, but it warms up rapidly, and at one point in the afternoon it gets so hot with the sun out that I am down to working in a tee shirt, and actually exchange my "sponge" stocking hat for a visor cap that says "El Jay" in bold gold letters—the name of a pump company here in town. I bought it at Goodwill. Brian and I have a few discussions about pumping and the pumper and the chief pumper.

We sheet off the second floor above the garage. The work swings. And we're happy in it for most of the day. The depression lifts. We're clipping along.

Bud is silent and sullen all day. If he says anything at all, it's a comment about how to do a job. I start to build onto this big resentment toward him, left over from the time a few days ago when he laughed at me after I notched out the stud (when he was the one who had called for it on layout!), but then toward the end of the day I have the most obvious insight, an insight I have lost track of for the past few days. Of course the guy is massively sullen and silent—he has this little baby with a big scar across the top of her skull, with this mysterious tumor growing on her spine, with the chemotherapy treatments every month, and God knows what sort of bills piling up at home. They are covered by insurance, but the insurance doesn't cover lost wages and traveling expenses and all that extra stuff. I have been thinking he just hates me, and I'm the kind of person that finds it hard just to let someone hate me. But now I realize that he's really depressed, not just moody—cosmically depressed, and nothing we can say or do will change that in any way. Even hearing Brian and me going on and on about whatever we go on and on about must piss him off.

Vern seems as steady as an old river, flowing along, flowing along, building the house. Even those outbursts of temper where he starts beating on things with his hammer seem to be all part of his flow, a flow that has been flowing the same way for years, decades even.

Laying the plywood is easy and fun and I have a few moments during the day where I think this is the job I most like to do in all the world. My body is still achy, my back tweaked out in certain positions, my legs strain as I bend over to pick up heavy lumber, and my left arm seems about half as strong as it should be, as if there is some malfunction in the muscle reflexes. I keep thinking I need to go to Steve, our chiropractor, and get myself adjusted back into some kind of order. At the least, I could do yoga, but I don't even do that.

The treat for the day is the arrival of Brian's entire family at lunch-break. It's his Dad's birthday and they're out cruising, so they've all come by to see what Brian Jr. is building. I meet his Dad and Mom, his wife Jennifer (again), his two sisters, and his Grandpa. His Grandma stays in the car with one of the grandkids. Grandpa is a trip, going right off into his old stories about his caddying days in upstate New York at a golf course where he used to make extra money ("sometimes fifty bucks a day") retrieving golf balls out of the duck pond on the course.

Brian seems happy, even happier than usual, and we have a great afternoon, even at one point when we are both up on ladders on opposite sides of a wall, up at least fifteen feet, and I start to lose my handhold, start falling backwards, and in grabbing back at the wall hit him in the head with the Sawzall; even then we still laugh and feel immortal and strong. When a little rain starts to fall at the end of the day and we are cleaning up, we lift our faces to the sky and smile. It's the spirit of Tommy Tyvek and we're sailing along under a favorable wind.

January 26

Thursday, and I can already feel the weekend approaching; there's a heightened awareness of time, of leisure, of possibility, of the fact that yes, there are other things in life, outside the bounds of this routine, this time-vice we operate in. My time is so strictly paced and every morning there's the same sense of the countdown. I've been getting up at six. Coffee and two pieces of toast with butter and jelly. I'm up alone, and while Kathleen and Shannon sleep, I read the newspaper and then get moving, filling my thermos with coffee and making my lunch, which I pack into one of those getaway mini-coolers, this one red and white and labeled, on the outside, "Little Playmate by Igloo," a kind of wussy name for a construction-worker lunch-bucket, but Bud has the same one and on any given lunch-break at any given construction site around the country, you would probably see at least half the workers with these same "Little Playmates." What is this? I mean, doesn't "Little Playmate" sound like something for preschool? Whatever—the thing works great. I put a container of blue ice along one side, then my blue enamel cup and a recycled Dijon mustard jar with milk. After years of funky-smelling thermos bottles, I learned to leave the milk out of the coffee and mix it at the site. Then my sandwich: peanut butter and jelly on sourdough. Then the extras: corn-chips, raisins, carrots, Fig Newtons, saltines, cubes of Swiss cheese, a banana, and an apple. I won't eat it all, but there's something reassuring about having it there with me.

I try to leave the house by 7:40, but today I'm a bit late, it's 7:49, so I know I'll get to the site a few minutes late. Still, I know the overall score, I know that twice this week I've been at the site ahead of Vern, and the third time I cruised in right on his tail. So I'm due a few minutes of slack. Not that anyone's really keeping score—except me and maybe Brian. Even though I sometimes suspect Vern keeps an exact tally in his book, he has never said anything to me about being either early or late. Not like old Bill Graziano in San Francisco, who used to say "Good afternoon, Joe," if I showed up even five minutes late in the morning. Bud is the last to arrive, nine times out of ten, but is this because of Carol Anne?

I take exactly the same route through town every day and space out on the drive. We have been talking about "one-pointedness" in my meditation class, the attempt to do one thing at a time with full awareness, so I am trying not to listen to the radio. Instead, I concentrate on the drive. The same streets, the same turns, the same line-up of single commuters in their cars, me one of them.

When I get to the site, Vern is already parked, backed in up on the gravel with his big, white, brand-new Ford truck. I start to roll out the tools; we never say much. The clouds are dark and threatening but the air is warmer than it has been, so we share a few very general comments about the weather. Vern's speech is hard for me to understand; he talks in a kind of contained drawl and keeps his words to a minimum. After working for him for five months I am beginning to know the signals. But I still have to ask him at least half the time to repeat what he just said.

I am, in fact, almost completely deaf in my right ear, but have full, normal hearing in my left. The condition goes back to early childhood. Today I hear my name being called and have no idea where it's coming from. It's Brian. I finally find him down on one side of the building holding up a three-quarter-inch piece of plywood for me to pull up onto the second floor.

At coffee-break, we sit under the new plywood of the second floor and stay dry out of the light drizzle that has started to fall. The discussion turns to roofs and falling off. A 12:12 roof is the steepest anyone has worked. That means the roof drops a foot for every foot it goes out. The steepest roof at the last house was a 9:12. The roof on this house is a 7:12, which isn't too bad. But it's going to be solid-sheeted, which is slipperier.

Vern tells a story about a 10:12 roof up the King River with a three-story drop-off. "We got out to site one morning and there was frost all over everything. The roofers came and I told them to be careful up there and they said, 'Nah, we're fine,' and about a half-hour later we're looking right out the window and here comes this guy, whoop, right past us, just slid right off." He's laughing when he tells this, like it was Buster Keaton or the Three Stooges, and I'm thinking three stories, Jesus, what about the guy? "Just busted up his ankle a little," Vern says.

We sheet most of the second floor, fighting the rain and trying to get the glue down onto the wet joists so the floor won't squeak. Brian and I work in quick, focused coordination, and we feel like we've done a lot by the end of the day. Tomorrow we should start on the second-floor walls. Brian cooks two elk steaks on his truck-bed barbecue for lunch.

Bud is as sullen and uncommunicative as ever. But when he makes a couple of ridiculous and obvious suggestions about my work, I shine it on. I'm in too good a mood today. I'm wearing my old seventies disco jacket and I'm stayin' alive.

January 27

TGIF. Another morning of warm, damp air, but no rain from the sky and for that we are thankful. After the initial tool preparation, Vern sets Brian to cutting fifty studs on the radial arm saw and takes me up top to start nailing together the second-story walls. I feel like I've been picked for the better job, so I'm excited. And nervous at the same time, wanting to do everything fast and perfect, wanting to prove myself, wanting to show Vern how good I really am. Bud is not here yet, so I feel like the favored son. Vern carries such a patriarchal aura, this feeling is not at all unrealistic. We are his three sons—he's our Noah. The Chin house is the ark. It rains all the time anyway so what's the difference? Still, I feel I'm struggling constantly for the withheld approval of the father. I have been working for Vern for five months, and he has only given me positive feedback once in all that time. One time, when he had me frame out a fairly complex, angled, two-ceiling box for one of the skylights at the Baxter house, I did it in record time, designed it and built it myself, and it looked good, it really did. He came by and without hardly looking or even really breaking stride, from the floor to me eight feet above on the ladder, just barely audible, like everything he ever says, he did say, "Looks good," and I believe that's the only time he has ever praised me. Maybe that's his strategy. We're like puppies running around under his feet, hoping for a pat, a scritch, a wink, any acknowledgment. And so we try. And so today I work my ass off to get this first wall just right.

Brian is down below, and in a break from stud cutting I have him cut the trimmers and headers for the window that is in the center of this wall I'm building. The window is four foot high and the top is supposed to be six feet ten inches from the floor. But we're not using continuous trimmers. So I have to measure up from the bottom of the bottom plate six feet ten inches, make a mark, then measure down from that mark forty-eight inches to make my mark for the top of the sill. An inch and a half below that is another mark, and the distance from that mark to the top of the bottom plate is the measurement for the lower trimmers and studs. I yell down all these numbers to Brian, who cuts them as fast as I call them out. We're flying. Brian and I, when we're into it, are all asses and elbows, we're balls to the wall, we're kickin' butt, we're shreddin' it. He throws all the blocks and cut lumber up to me and I lay it all in place and nail it off with the nail-gun. Except that when I get to the last point and I'm ready to nail in the header, I see it's all an inch too high. I've fucked up! The old inch-worm got me! Brian sees it too. "The old inch-worm got ya," he says. And I'm sick because Vern is right there too, the way he always seems to be when there's a fuck-up, and he sees the full horror of it. And Bud who is the favored, real, natural, first born, birth son of the patriarch sees it too, and everyone hangs their head to avoid my gaze because I have fallen from grace.

This is how the inch-worm does his treacherous work: When you have a line out in the middle of a board and you need to measure from it, you can either set a nail and pull off that, or you can lay your tape out on the board and set it at the one-inch mark. The reason for doing this is that the little half-inch clip at the end of every tape that hooks around the ends of boards is rather inaccurate out in the field. So the one-inch mark is more reliable. The problem is that in your measurement you have to add an inch to make it all come out right. I was measuring down on the trimmer forty-eight inches. But I made my mark at the number 48, and I was already up one inch on the one-inch line. With my faulty mark, the real, measured distance was only forty-seven inches. I should have been making my mark at 49—a one-inch fuck-up. The fucking inch-worm.

I correct the error in about ten minutes, using the big crow bar, pulling the wall apart, cutting the short trimmers and studs correctly, and nailing it all back together again. But the psychological damage has been done and I walk around for a few hours feeling grim and unworthy.

It's Friday, though, and after lunch no one can stay depressed for too long. Everyone's mood shifts for the better, even Bud's; even when it starts raining, we stay out of our raingear and let ourselves get wet, because we know in a few hours it will all be over for the week.

At the very end of the day, for some reason, the subject of Ernest movies comes up: *Ernest Saves Christmas*, *Ernest Goes to Jail*, and so on. It turns out everyone, even Vern, loves Ernest movies. These are movies about the dumbest fuck-up who ever walked the face of the earth, Ernest P. Worrall, who gets into all sorts of difficulties through the agency of his own stupidity, but who somehow, because he has a kind and generous heart, always triumphs in the end. I love Ernest movies, myself, but I have almost never met anyone else who does. A few years ago when the new Ernest movie came, Shannon and I raced off to see it one night. Under my persistent influence, Shannon has also become a big fan of Ernest. Kathleen refused to go. There was only one other couple in the theater, and before the movie started we spent about ten minutes comparing notes with them on our favorite scenes from the Ernest movies of the past. Now Vern, Bud, and Brian start to scramble over each other to describe their favorite scenes—the one where he's chewing his pen in the jury box and the ink comes out all over his mouth, the one with the rattlesnakes, the one where he's cleaning out the toilets at the camp. It's the closest we've ever come to a state of general hilarity during work hours and we all suddenly catch ourselves, stop and go back about our business, as if we had just been caught in some sort of bizarre comedic anti-universe.

The last hour of the day drags by minute by slow, tedious minute, because it is the last hour of the week and you wish suddenly that it would all be over and you could go home. When I do get home, I plop into the first chair I see in the overheated front room, and it feels like my arms and legs can hardly move, and it feels better than at any time during the rest of the week, because it's the exact moment before you have spent any of the long hours of freedom that you have spreading out before you. TGIF again and Amen.

February 1

WEDNESDAY. In an effort to recover from an arm injury, I've missed work the past two days. This is a problem I've been having with my left arm for a while now, for perhaps two or three weeks, since we began the heavy-lifting phase of the framing process. I've noticed considerable stiffness and soreness around the bicep, restricted motion, and, most important, a distinct loss of strength. At the end of the week, last week, I was having trouble lifting single sheets of plywood. Then on Sunday night, during a hockey game, I fell on my left side with my left arm extended and the pain was excruciating. It felt like the whole arm went numb. I called Vern around 9:30 Sunday night, after I got back from hockey, to tell him I wanted to go in to see my chiropractor about it, and that I wouldn't be in Monday morning. "Do you ever get anything like that?" I asked, trying to draw his sympathies into my situation. "All the time," he said, "all the time," implying, of course, that he shows up every day at the job and plays hurt, no matter what. He succeeded in making me feel like a creep for ditching work, but on the other hand I had my arm to consider. Two days of visits to the chiropractor, heat, Advil, and stretching out the shoulder, and it feels a lot better.

Brian greets me this morning with, "How was your vacation?"

And I do feel like I've been loafing around, watching the O.J. murder trial, and avoiding work because I can't take it, I'm not man enough, I'm afraid of the heights. We're really getting up there now. In the two days I've been gone, they've placed most of the roof trusses over the eastern end of the house. These are huge, factory-constructed units that combine the rafter, the ceiling joist, and all the interior bracing and collar ties. It looks impressive. Today there is an absolute downpour all morning, it seems to rain three inches at least, and I am soaked to the bone, even through all my raingear, by 10:00.

We take an extra-long coffee-break and while Bud and Vern sit in Vern's truck with the heater going, Brian and I sit under the plywood lid in the garage and talk about the job. Mostly we talk about what a jerk Bud is, although I try to say that he's such a jerk because of his kid's medical problems and that he's depressed and just takes it out on us. Brian doesn't agree. He seems to think Bud is just an inveterate asshole. But we're all right with our disagreement too, and it seems to bring us closer together as the two non-family members of the crew. Apparently Brian and Bud had been talking about me during my absence, and Bud had been putting me down about my skill level and Brian said he'd been sticking up for me.

But it's true enough now—I feel more and more like the geek on the job. And today with my recovering arm, my soaking-wet clothes, and my boots that feel full of water and like two giant sea sponges on the ends of my legs, I do feel awkward and ineffective and reluctant to get up too high into the structures above the second-story ceiling line. We frame walls all day on the second floor, run trusses over part of what we have framed, and stay close and busy. More and more Brian seems to take the harder and higher tasks as we work together, and I let him have them.

I stick with my mantra—*om shanti*—all peace. *Om* is the ancient word that is supposed to contain all sound, beginning as it does at the back of the throat and moving toward the lips as they close together at the end on the "m" sound. In meditation class Monday night, a woman handed my friend Greg a paperback book about two hundred pages long that had all the different mantras listed, something like the kind of book you can get as a guide for all possible babies' names. It gives you the mantra and then a brief description. With my eyes closed and never having looked in the book, I said, "Okay, now I'm going to pick my mantra," and I flipped through the book and landed with my finger in the middle of a random page. My finger landed exactly on the words "*om shanti.*" This has, in fact, been my mantra since I went on a ten-day silent yoga retreat with Swami Satchidananda in the early 1970s. What are the chances of my landing exactly on that page and that paragraph? It's impossible. Nola, the meditation teacher, looked at me with her fingers outstretched and a big smile and made that sound of the *Twilight Zone* theme music.

Still, if I keep saying this mantra and keep working and stay close to Brian everything seems okay. Vern is off in his own world, fussing and fuming half the time because things don't fit together, and Bud stays below, cutting boards for us on the radial arm saw and saying nothing except on rare occasions when he appears where we are working like some ill wind, stares over our kneeling forms, and offers suggestions on other, presumably, at least according to him, better ways to do what we are doing.

February 2

I'm still feeling a little behind the eight ball, particularly this morning. Brian takes charge and I follow around, do what he tells me, try to stay busy. The day is cool and foggy at first and then warm with a report on the radio of a large, orange, glowing UFO in the sky—the sun. An Oregon joke. And really it does feel like a while since we worked in warm sunshine. I have left my long underwear at home and the unrestricted movement of my legs feels like a special pleasure. Also my arm problem seems much better. I stretch it out a few times during the day and feel more strength there than I have felt in weeks. Let's hear it for the chiropractors!

We are finishing up the last of the framing on the second floor and working high. Framing and then sheeting the back side of what will be about the highest peak in the house. At one point, Brian is dangling twenty-five feet in the air with one foot on a window ledge and the other floating free, somehow managing to nail the far corner of the highest sheet of plywood. I feel an almost nauseated dizziness at the sight of him, and Bud yells up from below, telling him not to take such risks. A few minutes later, Vern shows up with one of the safety ropes. Brian is against the use of safety ropes, ever. As a former roofer, he feels he's invulnerable to injury from a fall. He's only nineteen years old and his blood pumps with that brash, immortal pride of youth. He's also six foot four, 240 pounds, and his feet in their giant, yellow rain boots are over fifteen inches long. Still, he claims he's part monkey and grumbles about having to wear the rope. "Vern was up there walking along that plate," he says, motioning to the top of a two-by-six wall that is not much lower than the wall we are working on, and he's got a point. I've seen Vern working in all kinds of precarious spots and I've never seen him put on the safety rope. I used it all the time when were up on the high, unprotected back side of the last house, and was glad to have it.

You put the thick, padded canvas belt around your midsection and then pay out the rope through a kind of climber's grommet that allows you to repel down the wall, although I have never done that. Mostly I tie off at the distance I want with a figure-eight knot, then work as much as I can with that amount of slack, and then retie when I move. I try to tie off at a rope length that would leave me dangling halfway between the roof edge and the ground. With that belt, which is not a through-the-legs harness, you would end up with an extremely sore ribcage if you did fall, but you would still be alive, at least. I was reading my Contractor's Board newsletter a few weeks ago, and it reported a series of deaths in Oregon, most of them involving roofers or carpenters falling off roofs or ladders and many of the falls no greater than twelve to fifteen feet, a level at which we not only operate regularly, but at which I personally sometimes take foolish risks. Above fifteen or twenty feet, the height gets my attention and I am always methodical and cautious.

Today, at the end of the day, we discover that the highest stud leading up to the peak of the highest wall on the west end of the house is significantly out of plumb. The top plate has to be removed, the stud cut down, and then beaten over to where it will be straight. Brian does all this standing with one foot on a two-by-six wall and the other on a six-by-eight beam. On the outside, he is a good twenty feet off the ground. I carry the tools and materials up and down the ladder while he pounds away, cuts, and saws. I would never have worked in that position. If I had to do what he is doing, I would (1) wear the safety rope, and (2) have a ladder with a much more accessible angle to the place that needs work. But Brian seems immune to fear, and chatters away while he hacks the offending stud into its proper place, sometimes balanced on the narrow board to cut the angled line above his head. Bud and Vern look on with doubt and concern as the operation proceeds.

It is the last thing we do for the day and when we are done it is already a little past quitting time. We race to pack up the tools and be gone. Vern says he has to meet with a supplier and leaves while we are still putting our own tools and belts away in the backs of our cars. I get home and feel good, like I have finally gotten back into the rhythm of the work—tired, but good.

February 3

Friday again already. It hardly feels like I earned the right to feel Friday feelings today because of my short week. I once had a job in Southern California where I worked only on Tuesdays and Thursdays. I would have the equivalent of Monday feelings and Friday feelings and no mid-week feelings at all. Of all the days of the week, it seems clear to me that Fridays are the strongest and most distinctive. Right from the opening bell, right from the moment we first roll out the cords, plug in the air compressor, hear its chattery, rhythmic song as it pumps its tanks up to pressure, there's a sense of excitement, of the end in sight, of the weekend looming, an anticipation of the taste of freedom. Freedom would have no significance at all to us if it wasn't for the long, hard, sometimes uncomfortable hours we spend with our time structured and locked down, with that vaguely oppressive and constant feeling that the boss is watching, that our performance is under scrutiny, that we have no right to chat or laugh or even talk much in the serious arena of home construction. Nevertheless, Brian talks, talks, talks. It's a loaded situation financially, because we all know we are working against time. Every second that Vern pays us for that we are not actively engaged in driving a nail, cutting a board, snapping a line, laying out a wall, or whatever else we might have to do in the direct service toward completion of this project, is money that has been in a sense stolen from him.

I wonder sometimes if this is the reason for Bud's surly, random acts of pettiness and micro-management. At the end of the day yesterday, Bud laid up a single small piece of plywood. It was eight feet long and about twelve inches wide, and he had a little trouble doing it alone, but he would rather shoot himself in the foot with the nail-gun than ask for help. Brian and I both noticed that he had put it up backward, that is, with the printing facing out. The other side has pre-rendered lines for the studs and is therefore properly the side that should be facing out. Brian and

I kept this observation to ourselves, with a kind of secret satisfaction that we had caught Bud, the taskmaster himself, in an error. But today, about an hour into work, Brian cannot restrain himself from saying something. I am down on the ground cutting blocks on the radial arm saw when I hear him telling Bud that his piece of plywood is in backward. A discussion ensues in which Vern, who has been drawn in as the authority, basically says that although it is better to put the lined side facing out, it is not imperative. Bud has been technically caught in a screwup, but it is not a fatal one—the board will not have to be taken down and changed.

The consequences are inevitable. Within fifteen minutes, Bud is up on a ladder above a big double-door header we have installed, claiming that the blocks have been put in wrong. It seems like a direct act of retaliation. These are three-and-a-half-inch blocks that are as close to meaningless as any part of the framing could be. More than likely they will never get even one nail or screw, but now, because they are fractions of an inch off a perfect sixteen-inch-on-center layout, Bud fusses and beats at them and wonders aloud who is responsible. It's Brian, of course, who put them up in his practical, down and dirty way, fast and sufficient, but, as Bud has discovered, slightly incorrectly. Brian smiles at me and under his breath says, "Is my analysis of this situation correct?" "He's getting you back," I say. "I know," he says, and we look at each other and understand that in the constant battle over who makes what mistakes, we will always come up on the short end. The situation is governed by that age-old line that separates and always will separate management and labor. Brian and I are labor, Bud and Vern are management. We are the noncoms, they are the coms. We are the grunts and they have the maps and the compasses, and in the end they give the orders and we follow.

In the afternoon, I work mostly alone, nailing off all the

sheeting that we tacked up yesterday. These are four-by-eight-foot sheets of particleboard that we set into place and held with four or five nails. Now they must be nailed off in a pattern in some places as close as every two inches along the edges and every six inches in the "field," that is, out in the middle of the sheet along the studs. All the sheets are up high, and I am now operating off the big extension ladder, fifteen to twenty feet in the air. I'm in my alert mode. At first, I can feel the clutch of fear every time I climb up. I have to haul the nail-gun with me, extending its long, heavy umbilical of air hose behind. I have to set the base of the ladder onto scraps of wood that I set down onto the surface of the wet, gooey mud that surrounds the building. I sometimes feel that the ladder is in danger of moving off sideways across the slippery surface of the plywood wall, and at other times, when I am working above my head, on tiptoes on the highest rung I dare stand on, pressed against the face of the building with nothing to grab onto, I feel that with one awkward or sudden move I would fall back and land head-first on the wet ground below. I think of the pain, I think of death, I think of Kathleen and Shannon and Emily, and then I start saying my mantra again and decide that if I die I will die in a state of spiritual if not physical balance, and this thought cheers me. I become more adept at nailing. I find it easier to move the huge ladder from position to position. I start to lose the resentment I have toward Brian, who seems to be the one who gave me this job, even though I make $3.00 an hour more than he does. I start to feel satisfaction in the perfect, smooth, nailed surfaces I am leaving behind me. I remember how much I like to work alone and in silence. The last hours of the week seem to accelerate toward their conclusion.

And finally, I go home with my January paycheck in my pocket and feel the sweet luxury of free time seep into my being as I sit in our front room and drink coffee and contemplate the precious weekend ahead.

February 6

MONDAY morning, Monday blues. Except for some reason I don't feel too depressed this morning. The morning is cool and foggy but not cold and no rain is falling. Nor is there any rain in the forecast. In fact, on the Weather Channel last night I saw a graphic that showed sun for Tuesday, Wednesday, and Thursday.

Brian had gone off to Portland for the weekend with his wife and is full of the story when he arrives a few minutes late. He seems buoyant and jovial through the first part of the day. And we make one of the all-time spectacular throw-and-catch moves. I am up on the outer edge of a second-story shed-dormer window, nailing in frieze blocks between the rafter tails. I drop one of the blocks and it falls through the roof framing and the ceiling framing and down into the garage. Brian, who has been cutting for me, climbs down to retrieve it. He makes a motion as if to throw and I say, "No way in a million years." It's a two-by-six block about ten inches long. He says, "We've got to try it, it's one of the big events in the Carpenters' Olympics." This is an ongoing fantasy we have about a Carpenters' Olympics that would include events such as behind-the-back plywood cutting, nailing for speed, cord throwing, and this event, block throwing from two stories down through tight framing. He's right—we have to try. He sets and throws the block perfectly up through a sixteen-inch opening in the ceiling, and then up past a header that marks the base of the dormer, another gap of about sixteen inches, and right up into my outstretched hand. Even better, I catch it perfectly—no double clutch, no fumble, no nothing. It seems an omen of a good day and we breeze along, setting rafters for the next hour.

But then Brian and Bud get into an abstract discussion of safety and fear. Brian basically says he doesn't mind taking risks, that he's not afraid of death and doesn't want to become one of those people who are constantly worried about safety. I tell Brian that it's his youthful sense of immortality. Bud says that he's just big and stupid. If there's one thing I've learned about Brian, it's that you never, ever, call him stupid. I suspect he was called stupid in school a few times, possibly teased, and though he's not stupid, I can imagine that there were times when he put very little effort into school work. Plus he looks stupid—big, oafish, goofy, Alfred E. Newman on steroids. Anyway. Bud's remark causes him to smolder and fume for the rest of the day.

I'm working with him and I get the brunt of it. He's definitely gone into his Bad Brian mode. Bad Brian is one of Brian's alter egos, a Brian who takes no initiative, who tends to get perched up somewhere in the rafters waiting for work to come to him, whose mind is far, far away from the task at hand, who is depressed and pouty, who makes mistakes and refuses to acknowledge them, and who crows when anyone else makes a mistake, particularly me. The Bad Brian is a very hard person to work with. Still, we struggle along through the day. At one point I manage to get a smile out of him by singing "The Froggy He Am a Queer Bird," but it is only a temporary relief.

The overriding fact remains—no matter what peoples' moods and attitudes, the boards continue to go up into place. By the end of the day, we have most of the south half of the second-story roof framed. This is by far the most complicated roof in the house and it is a piece of cake compared to the eleven-gable roof design on the Baxter house. Vern, old man river, just keeps rolling along, lugging up the sixteen-foot pond-dried two-by-tens with never a word of complaint.

February 7

A Tuesday. I get to work late, and Vern has already rolled out half the cords and seems to have a bereft, abandoned air, his body and face as always communicating far more than his mouth. If his expression could speak, it would say to me, "Here I am, have to build this damn house all by myself, I don't know why I even do this, I don't get paid, all I get is what's left at the end . . ."

The day feels difficult all along. I'm having problems with Brian again. He came in about ten minutes after me with his Bad Brian face on and said he'd been up again past midnight, this time mediating a fight between his wife and his mother over a sick cat and whether it should be allowed to stay in the house. Brian and his wife are living with Brian's parents while they try to save enough money to buy a house of their own. Tired and grumpy, the Bad Brian makes faces, drags his heels, fouls up on his work and claims it isn't his fault, counters my every suggestion, and crows louder than ever if he discovers any error in my work.

I hardly know what to do about it, except to try again and again to stay away from the emotions and keep busy. Say my mantra as much as possible. Don't engage him in conversation. There's no real winning. If you want to be in synch with him you have to become a worker like the Bad Brian worker, slack off, talk about irrelevant things, sing stupid songs, and go at a snail's pace. Which I hate. I have to stay busy to stay sane.

I have a constant anxiety that Vern and Bud are watching me from wherever they're working; today they are on the northeast side of the house running the last of the roof rafters while Brian and I finish framing and sheeting the two small shed dormers on the southeast side. They're almost half the house away, but still, I'm sure they're keeping track of what I'm doing and how I'm doing it. Because I'm older and get paid more, I feel that I am ultimately responsible for everything Brian and I do together.

And maybe there's a hidden irritant here—because Brian told me a couple of days ago that he had asked for and finally been given a $1.00-an-hour raise. And I'm thinking—where the hell's my raise? I feel like I'm constantly sabotaged and made to look inept and that the dollar he got was somehow at my expense—that is, if he gets a raise and I don't, how is that supposed to make me fuckin' feel? The whole thing makes me sick on some level and on a Tuesday afternoon, even with the sun out and a forecast for bright, clear, warm, sunny days ahead, I hate work, I hate Brian, I hate the whole mess, and just want to go off somewhere and lie in bed and eat crackers and watch television. Maybe it's a Tuesday syndrome. When you spend eight hours a day walking around with somebody hovering on the outskirts of the scene watching your every step, judging it, evaluating it, calculating its very worth, it cannot help but make you feel trapped.

At the very end of the day, I manage to sneak away from Brian and work for a while alone, cutting baffles to deflect the blown insulation away from the vent holes between the rafters. This is fairly uncomplicated production work and I am happy to be alone. As someone once said to me a long time ago when I first began working at hourly wage jobs—"It's not the work that gets you, it's the people." How true that is! Today I even dread coffee-break and lunch because I feel like I have run out of things to say. Vern passes around some innocuous little conservative handbill that promotes capitalism with a silly parable about some cows, but I can almost not stomach it, and later I lie to Brian and tell him my father was a card-carrying communist.

Maybe all the tension on this job site is political. We have to listen to Rush Limbaugh again for a while on the radio, and at lunch we have a brief discussion of the minimum-wage law with everyone but me agreeing that the only reason anyone ever has less money than they need is because they're lazy. Today the whole thing makes me want to go off somewhere and crawl into a hole. And where are my prayers and my mantra and my higher vision of the great and inherent goodness of humanity? Out the rough opening of the upstairs window in a cloud of sawdust.

February 8

A really difficult day. The bright sun that comes up and beats down through the fog starting around 10:00 does nothing to brighten my mood. Actually, at first I am fine. I am working alone, putting in vents and baffles between the truss joists in the main part of the house. Brian is alone, working on a wall up in the second floor, a project he started yesterday. I can't help wishing him ill. I know it's childish, it's ridiculous, and every negative thought I have will come back to haunt me, but I still can't help it. When I start to hear the screeching sounds of nails being pulled out—a sure indicator of a fuck-up in progress—I'm filled with happiness, and when he asks me to get him the big crowbar, it can only mean one thing—that his fuck-up is so monumental it now requires a major extraction. I never go to look at what he is doing; I can see, however, that while I am breezing along getting everything perfect, he is living through a narrative of bent nails, missed cuts, split wood, and reverse construction.

Later in the day, we are working together, and as usual he wants to take control and manage it all. When he screws up, as he does when he attempts to get the roof sheeting started, and I intervene, he hates me, and becomes passive-aggressive again, sulking and unable to lift a finger to help me at all. The problem, I guess, is that I see it all as my responsibility. And I get nothing from Vern and Bud but criticism. So I'm stuck in an impossible dilemma. All I can do is keep working, keep banging away, and above all keep reminding myself that it's only a job.

Right after lunch I drop the Sawzall on my head from a ledge nine feet up. I have it balanced on the top plate below the roof line, where I have been using it to cut away a portion of one of the frieze blocks. While moving my ladder, I accidentally catch the power cord, and the Sawzall comes down right onto the top of my head. It's a heavy tool with a fair-sized housing and a long reciprocating sawblade. I see stars for a while and have a colossal headache for the rest of the day.

In the middle of the night I wake up thinking about how much I hate working with Brian and also that I am about to have a brain aneurysm and die. In my panic, I cannot help feeling the sadness that my last day on earth should have included such a disappointing eight hours of discord and despair. But I don't die. Instead I drift back to sleep, and wait for the 6:00 alarm.

February 9

I wake up this morning late and full of dread. My head still hurts, but really, I think my main problem is that we're beginning to work our way higher up into the building. Sure, I'm worried about my relationship with my co-workers, particularly Brian, but really, I think what's bothering me most is that I have to get back up onto that god-damned roof again today. The issue with Brian and his control freakism certainly doesn't help matters. He has to do it his way.

And I'm telling all this to Kathleen this morning, and she just says "easy does it" to me, that's the totality of her advice, but I think, as I get there, as I show up at the site, that's right! That's the key. I *can* easy does it. This is just a job, for Christ's sake. As the Russians say, "My job is not a wolf, it's not going to run away into the forest!" Fear and anxiety have turned me into this super-compulsive carpenter/worker, worried that they don't like me, that they think I'm a fuck-up, and it's crazy. What do I care? It's just a lousy $10.00 an hour job and you'd think I was in charge of negotiating an end to the war in Bosnia.

So today I simply show up and act as Brian's gopher and go purposefully slow and let the work evolve. And as the day goes by, it gets better. I let Brian sheet one entire side of the big roof, twenty-five sheets of plywood at least, and I cut for him and put in the lookouts, and for most of the morning I stay on the ground and stare out at the golf course and chat a little with Bud, who seems friendlier then I've ever seen him. Maybe he was picking up the vibes from yesterday. I'm alone a lot of the day, and that seems fine.

At coffee-break, Brian brings pictures of his elk hunt, grotesque snapshots of him perched beside this huge, dead cow elk with blood all over the snow and his gun against the carcass. He's in his yellow rain pants and a tee shirt and giving the old thumbs up sign and it makes me want to turn vegetarian. But I tell him how great it is anyway. Today he's being the Good Brian and he's fun to be with and he hardly gets on my nerves at all.

In the afternoon, I take the nail-gun up on the roof and shoot off the entire thing on both sides. Six inches on the edges, twelve inches in the field. I like being up there. I have no fear, or almost no fear, and walk around easily. The weather is dry and clear and the footing is secure, though my back gets out of whack after a while, and I have to take off my nail-belt because it feels so heavy. The hardest part is walking with my feet constantly bent at a thirty-degree angle. I'm like some Himalayan goat. I try to vary my position, first parallel and then perpendicular to the rafters, but in the end both positions are uncomfortable. The gun works well, blam, blam, blam, and I go along, alone, quiet except for the percussion of the air with each staple that it fires. The staples are big, about two inches long, and shoot into the wood easily with the compressor set at one hundred psi. The sun is out, the sky is a bright blue, I'm down to my tee shirt and dripping sweat. I've finally taken off my thermal winter boots and I'm wearing my high-top sneakers for the first time in months. This roof, by far the biggest section of roof in the house, the part of the roof that covers the large livingroom, the entryway, the master bedroom, and the big second bedroom, is entirely sheeted and nailed off by the end of the day. Bud and Brian are starting with the soffit plywood, following behind Vern, who is hanging the fascia boards and barge rafters.

I feel relieved when 4:30 rolls around, tired but happy that it's Thursday. "Tomorrow it's TGIF," I tell Brian, and he says, "TGTFO." I say, "What's that mean?" and he says, "Thank God today's fucking over."

The forecast is calling for more sun tomorrow and then the Arctic Express moving in over the weekend. A lot of talk today about how we could be shoveling snow off the plywood come next week. Looking out over the green, sunny, golf course, the golfers coming by in a steady parade, in shorts and tee shirts, the soft, fragrant air of early spring wafting around, mingling with the smells of new wood, oil, nails, chalk, and leather, it seems hard to believe that there is any winter left anywhere on Earth. During the ride home, on the radio I hear the commuter personalities talking about the unreliability of the Groundhog Day prediction. Punxsutawney Phil says six more weeks of winter.

February 10

I spend the first two hours of the day in the dentist chair getting a temporary cap put on my broken tooth. The time in the chair feels out of place, an interruption in the rhythm of the work week. For a while I am waiting for the glue to dry on one stage of the operation, and I can see through the office to another dental chair where a man is being fitted for dentures. He is a skinny man wearing jeans and cowboy boots and I think, from what I can hear in the distance of his conversation with the dentist, that he operates heavy equipment of some kind in the logging industry. He almost chokes several times while the fitting is going on; apparently the casting is being done deep in the back of his mouth near his throat, and I see him cry after one episode— it is like watching someone being subtly tortured. And then, only minutes later, I am squirming and clenching as my dentist hacks away at my gum line in an effort to clear away excess glue from the post he has just fitted. I try to lose myself in the pain, go into it, repeat my mantra again and again as if it could save me from anything.

When I get back to work, they are just finishing up coffee-break, and they ask me about my vacation, about my secret date, about all the slacking off I have been doing in the first part of the day. "Just kidding," Brian says, the punctuation that he leaves after every jab, every insinuation. "Don't you ever kid in the opposite direction?" I ask, but in its own way that would probably be just as subtle and destructive.

For two hours, Brian and I cut and fit the rafter tails onto the trusses that span the small roof above the diningroom area. The work is simple and direct. When the tails are in place, we cut and place the vents and the frieze blocks. The work ends up looking okay. Vern and Bud are working to finish up the soffits and fascias on the big roof at the east end of the house. The weather is alternately cloudy and cold and sunny and hot. I take my sweater and jacket off, my hat, my gloves, and then start to freeze and load it all on again.

At lunch, we sit in the sun on a stack of plywood, and I listen to more elk- and deer-hunting stories—the elks that got away, the wounded ones that lay still behind a tree, that played possum, the butchering, the luxurious accommodations at the campsite where Vern and Bud go hunting, the efficiency of Brian's site. I am the audience for all of this, as they try to top each other, reaching back into their collective memories: who let the biggest ones go, who went on the steeper and more rugged roads, who walked for miles at the age of thirteen carrying an elk head by its ears, who fired and missed from over five hundred yards away, at a four-point bull that never flinched?

After lunch, we go back up onto the big roof and roll out black paper. Brian has been a roofer and takes charge, has all the techniques. I follow along, nailing, setting, trying to keep my grip on the black paper. It's exhausting to walk all afternoon with my feet at that awkward angle, my toes trying to clutch and dig in through the very soles of my boots. With the paper in place, there is less sense of security. At one point, I am standing at a spot with insufficient nailing and the paper suddenly rips away from under me. I sprawl down onto the roof but catch myself before I fall too close to the edge. Brian finds the sight extremely amusing and spends a few minutes trying to imitate my expression as I fell. I'm not sure I was in any real danger, but the experience is a warning, and I walk even more cautiously for the rest of the day; for half an hour my tongue tingles with that coppery taste of spent adrenaline. "I never trust paper," Vern says later as he edges down the roof toward the ladder.

There have been predictions all week of rain and snow and freezing temperatures over the weekend, so we are hurrying to get the roof completely covered. "Nobody leaves until it's all papered," Vern says. I know we can do it. We're working fast, nailing as little as possible. At 3:30, Bud and then Vern join in, and for the last hour of the week we are all four working together on the same project at the same time, an unusual gathering of our synchronized, collective labor. Only at concrete pours is there anything like this cohesion. A gentle, happy, Friday afternoon mood embraces us all, and the last hour seems full of easy talk and light laughter as we roll out sheet after sheet of black paper. Although Vern does say that he had hoped to have the roof of the entire house sheeted by the end of the week. "If Joe hadn't gone off to the dentist . . ." he says—more light ribbing, but I let it flow easy off my back, it means nothing.

I am the outsider here, and they have seen that I can take the jokes if I have to. I must seem a rather odd character to them; they can't quite fit me into the grid of their experiences and expectations. They don't have any way of pinpointing me demographically.

At the end of the day, the bottoms of my feet are burning with a kind of raw, generalized blister from gripping so hard on a slant. I go home and take a shower and crawl into bed. It feels like the softest, safest place on Earth, warm, comforting, protective. I fall asleep for an hour before basketball practice (I'm still coaching Shannon's team—we are 0–9 so far) and bask in the feeling of the weekend ahead. I feel ground down to raw nerves by the events of the week, still sore in my head from where the Sawzall hit me, sore in my left arm with the continuing tendonitis, sore in my hands from various cuts, one in particular on my left thumb where I have a deep cut from my tape measure, which has a barb sticking up at around the thirty-inch mark. On the same thumb, a small metal splinter from an eightpenny nail has embedded itself under the skin and is just now becoming infected. The soles of my feet are sore from gripping on the roof, and my whole body feels bent over and stiff and knotted up like a tangled rope that has been left out in the rain for days and then dried and hardened in a hot, blistering sun. It will take me all weekend to uncoil. I just want to sit by the fire for days and do nothing.

February 15

WEDNESDAY already. We've had two days off because of snow. It rarely snows here, but on Monday morning we woke up with two or three inches on the ground. Vern called to say we would wait a few hours to see what happened. By 10:30, a real mini-blizzard had developed and work was called off for the day. "It's not the kind of day you want to be up on a roof," Vern said to me over the phone. "Well, I'll see you tomorrow unless I hear from you," I said. But Tuesday morning I checked in with him at about 7:15 and there were five or six inches of new snow on the ground, and he said, "No use fightin' it," and we took another day off.

It was Valentine's Day and I spent it with Kathleen in front of the fire, eating, reading, napping. We made a snowman with Shannon in the afternoon. Shannon said it was the first snowman she had ever made. She grew up in L.A., and for years the only snow she ever saw was on Christmas cards. Kathleen grew up in Cuba, and she told me what it was like to see her first snowfall when she was sixteen years old and had just moved to Rye, New York. I grew up in New England, but I still get a thrill every time I'm around a good snowstorm.

Yesterday the landscape was beautiful and white but the temperature was climbing. By this morning, everything had turned to wet slush. There had been a forecast for more snow overnight but nothing materialized. Vern called at about 7:15 and said, "Well, I guess we'll go over and take a whack at her." In Vern's conversations, the house is always female. Or maybe it's just work in general, as in his questions to us when we have been on a project for a while: "How's she goin'?" or "That about got her?"

I'm in no mood to take a whack at her at all this morning. I'm all set to spend another day in front of the fire. Now, after the ill-fated phone call, I have to switch into work mode. I make my lunch, I gather up my tools, I make sure I have all my raingear and hats and gloves, and I show up at the job site about fifteen minutes later than everyone else.

It's about thirty-five degrees and overcast with a trace of drizzle falling. Vern and Brian are already busy shoveling the wet snow out of the house. The house is filled with snow except for the part we got roofed and sheeted and papered last week. We shovel out the rooms as best we can, and start sheeting the small roof section over the diningroom. Brian does most of it. When I get up on the roof, my wet rubber boots start to slip. We're about fifteen feet off the ground, but it feels like several hundred. Everywhere underfoot is slippery and unstable. Patches of snow and ice cover eighty percent of the site, even though things are slowly melting. Brian, the veteran roofer, just keeps working along, seemingly oblivious to the dangers he's flirting with. He's like one of the higher primates.

I go to get a safety rope and try to remember how to thread it through the rappelling buckle and tie a figure-eight knot. But by the time I get up on the roof, it's almost too late, so I spend the rest of the morning feeding Brian materials from *terra firma*. I feel like a wuss, but at the same time I hate the feeling when my boots start to slide out from under me on wet plywood or wet black paper loosely nailed. On a solid-sheeted roof you have nothing to grab onto. The wet paper is even slipperier then the wood, which at least has a little grain to it. None of this was a problem on Friday, when it was hot and dry and my shoes felt like solid magnets, like those anti-gravity grip-on shoes spacemen wear in old sci-fi movies. Also, on Friday, we were only eight feet off the ground. Here we're two stories up. The difference between maiming and death. Does it matter? The choice, in my mind, comes down to a wheelchair or a coffin.

I'm happy when we finish with the little diningroom roof and get onto the bigger roof section that extends from the south wall of the garage all the way up to the high point above the upstairs bedrooms. Now, with Brian on the roof and me cutting and feeding him plywood, we fly ahead. We're back into our rhythm and things are really cooking. Brian has brought his brand-new $250 boom box to the job, and during coffee-break and lunch he plays me his Ray Stevens CD, including the hit from yesteryear, "The Convention," a song I have been reciting lines from for years. "How'd you get that big Harley up there in the motel room in the first place?" and "Bubba—you are out of the shrine . . . that's right!" This CD somehow eases my relationship with Brian today.

I have started off feeling grouchy and cold and uncomfortable. I wear my jeans and rain pants even though not much rain falls. I have on a turtleneck, a sweater, and my heavy jacket, and I wear gloves through the first half of the day. The weather becomes more hospitable in the afternoon and the work goes well.

At the end of the day, there's a small snowball fight, with Brian and me teaming up against Vern and Bud. It's tentative and hesitant, particularly on my part. Brian seems to relish the opportunity to blast the boss and the boss's son with apparent impunity. We pack up the tools and leave and I'm feeling good. I'm the last to drive off. Half a mile away, I come upon Brian stopped by the side of the road with the hood of his truck up. He's scratching his head, and I immediately think, shit, his truck's broken down and now I'm gonna have to drive him all over the map looking for some stupid part when all I want to do is go home and crawl into the shower. Still, I'm his friend, I'm his co-worker—I have to stop. I slow down, come even with the front of his truck, roll down the passenger window, and I'm about to ask him what the problem is when he blasts me with a giant snowball. There's snow all over my front seat and all over me. He's back in his truck and making a fast getaway. I've been had—big time. Man oh man—I'll have to come up with some way to get revenge.

February 16

Today is my thirteenth AA anniversary. I've been clean and sober since February 16, 1982. Reagan was in his first term. Nobody had computers or CDs, O.J. was just getting started in his broadcast career, I was working for Paul Murphy, the "Bad Boy of White-Trash Architecture," remodeling a small house in Venice with astro-turf counters, boiler-plate floors, rippled plastic siding for interior walls, and barbed-wire ornamentation. Today I start the day off with an early morning AA meeting and show up at work just as Vern is driving in. The weather is cold and overcast but with no sign of rain. I wear my sneakers because Brian told me that sneakers grip better on a roof.

Brian and I start off sheeting the big roof above the garage. This roof, though it is big and steep, has two dormer windows that stick out about half-way up. They're shed dormers with al-most flat mini-roofs, and they provide a break from that long, sickening expanse of 7:12 plywood. Plus it's dry. And I have my sneakers on. For the first part of the day, I'm back and forth onto the roof with almost no anxiety. Next, we sheet a small, flat roof that sits above the staircase, and this one, too, seems easy, be-cause it's flat, even though it sits a good twenty feet off the ground on one side. The ground still has traces of snow but most of it has melted. On that back side of the house, the dirt is really thick, soft mud, and I try to comfort myself with the idea that even if I did fall twenty feet, I might only break my leg. As I stand around cutting sheets for Brian and sending them up to him, I try to figure out how fast I would be going when I hit the ground. At 32 feet per second for the first second, I would therefore travel 320 feet in ten seconds and 1,920 feet in a minute, which is less than half a mile. Sixty miles per hour is a mile a minute, so I would only be going about 20 miles per hour, which, if you think of a car accident, would not be that fast. For some reason this comforts me too, and I think, well, perhaps I'll break my leg and collect worker's comp and disability for six months and stay home by the fire and take it easy. All of this seems to be in the service of convincing myself not to be jumpy when I am up on these god-damned roofs.

But when the time comes to go up on the highest roof of all, one that is about twenty-five feet up and steep, I absolutely refuse to go on the one open side, and Brian does it all. I feel I have lost some status here, I feel wimped out, and can see in Brian a kind of satisfaction. He talks about some of the reckless roofers he has worked with in the past. But still, there he is, down at the very lip of this steep roof, laying out the paper and nailing it, with the only toe-hold an inch and a half of fascia board that sticks up above the level of the plywood. This has to be papered over, so the inch and a half becomes somewhat blunted. He sings and jokes and lets me watch his daredevil routine and it makes me dizzy even to look at him. The other side of this small roof is over a patio with only an eight- or nine-foot fall, and by compari-son that feels like nothing. I'm happy to work on that side.

At lunch, for some reason, the topic turns to gruesome acci-dents. Bud tells a story about the car wreck where the guy's head ended up in the glove compartment. Vern tells one about find-ing a dead driver at a car wreck whose jugular vein had been sliced going through the window. The way bodies fall apart after they are run over by trains is discussed. All of this while I'm eating my peanut butter and grape jelly sandwich.

Vern and Bud also want to hear the details of how Brian suckered me into the snowball attack yesterday, and they laugh through the telling. But Bud is eager, in the afternoon, to see me get my revenge. Toward the end of the afternoon, we are both down on the last roof that needs sheeting, the roof that faces the golf course. It's a big, steep roof with two huge skylight holes in the center of it, but at the outset we are down low. Bud comes around below me and throws me up a big, perfectly formed snowball. Brian is bent over with his butt in the air, nailing off the edge of the plywood. Without thinking, I throw it at the perfect target. Vern is down below him. I make a clean, bull's-eye hit, there's an explosion of snow, and Brian stands up fast, smiling and ready for his own counterattack. He loves snowball fights. But Vern gets totally pissed off and storms around yelling at everyone about "no horseplay on the job." I act all innocent like I don't have any idea where the snowball came from, and Vern does not single me out, whether he knows or not. And really, he's right—it was a dangerous thing to throw a snowball at Brian's butt while he was bent over the edge of the roof. Still, Bud and Brian and I smile at each other, the three sons now, bonded by our common crime, caught by the patriarch who storms off to cut another fascia board. There has been snowball fighting from time to time all day, and it seems clear that nothing has brought us all together like this snow, not even the Ernest movies.

At the very end of the day, when we have packed all the tools away, Brian comes from around a corner and throws a really hard snowball at my legs. It hits me and also partly hits Vern who is standing next to me. Apparently this is considered after-hours horseplay and we all leave laughing, knowing, I guess, that by tomorrow the snow will be gone.

February 17

Friday already and pouring rain. Right from the git-go. I get out of my car and put on my heavy sweater, rain pants, rain jacket, and Gilligan hat. No gloves, since the temperature already seems at least warmer than it has been the past few days. Besides, what would be the point? My hands will be soaked in ten minutes regardless. Vern is in full raingear too, but still wears his leather boots with the crepe soles, presumably because he will be up on the roof. And so will I.

Brian arrives, and we start in right where we left off yesterday, shortly after I hit him with the snow ball. There are lookouts to cut, some rafters to straighten, and then we are ready to start sheeting the last roof, the big roof with the giant skylight holes that will illuminate Beverly Chin's painting studio. The rain pelts down, but my sneakers seem to hold all right on the plywood. This is a big, steep roof, but it ends only ten feet off the mud, which seems survivable.

What is it that makes me so fearful? I've been reading Peter Matthiessen's book, *The Snow Leopard*, which includes some of his thoughts on the fear he feels as he climbs around the high, icy, Himalayan passes of Nepal, and he concludes that this fear of heights is a primal, clutching fear that represents the very basis of all attachment to life. I sometimes think it has to do with my feelings of responsibility to Kathleen and Shannon and Emily, but I think Matthiessen may be right—and isn't it really death we're talking about here? Still, I think, my fear can be unlearned to some extent: The longer I am on the roof, the more time I spend working on roofs in general, the more comfortable I will become, and this seems to be true as the day goes along. I try to pray, and repeat my mantra whenever I think of it. I hope, if I die, that I will die in a spiritual state and that this will ease my passage through the dark dimensions of the afterlife.

Halfway through the morning, we break off from sheeting and go back over to the garage roof to paper it. And this we do in the pouring rain. The black paper is pretty slick and slippery when wet, but I methodically lay up two-by-four toe boards as we go, something Brian dismisses as chicken shit. He rides me a little. "Did you leave your guts at home?" But I have to remember— he's a nineteen-year-old immortal god, he's worked as a roofer on summer jobs since he was fifteen, and he's a cross between an aerial thrill seeker and a higher primate. I'm fifty-one years old, afraid of heights, brittle, somewhat restricted by shoulder, wrist, and back injuries, and in no way should I even consider trying to keep up with his pace. Once the toe boards are nailed down, I can work at my own pace, and in some ways, I am almost as fast. Brian at one point seems to resent this slightly and we switch roles. He's cutting for me, he's down and inside, and I'm up on the roof laying down the paper.

In the afternoon, we go back over to the studio roof. Here, it's a little more hairy because of the big skylight holes, holes that are six feet by eight feet, plenty big enough for a body to fall through. At times, late in the afternoon, my feet slip a bit on the slick paper, and I have to catch myself, but not before I visualize the whole down-the-drain scenario as I slide across the slippery black paper and then through the skylight hole and into oblivion. In my worst-case scenario, I flip in mid-air and land either on my head, in which case I am dead, or on my back, in which case I am a paraplegic for the rest of my life. I wonder if Kathleen will still make love to me if my legs are paralyzed, and I have a fantasy of becoming an over-fifty wheelchair racer, perhaps doing the Boston marathon.

It rains all day and yet we keep working through it all, without let-up, almost against the clock and against the two lost snow days at the beginning of the week. Racing to have all the roofs fasciaed, soffitted, sheeted, and papered, we work a little past 4:30. And yet it is done. When I ask Vern, "Is that it for the roof?" he says, "Pretty much," and I think I will not have to be up there again, at least not for any extended stays. Perhaps I can defer to Brian for what few roof jobs remain. We do still have high wall sheeting of the gables to do, but this will be done from ladders, and for some reason, ladders feel more reliable to me. At least your feet are secure.

I get home wet and cold at about 5:00 and feel utterly exhausted. Shannon goes off to see the girls' high-school basketball game. Kathleen seems interested in perhaps going to a movie, but I am too tired even to think of it. All I want to do is lie on the couch.

About 8:00, I'm lying there on the couch, Kathleen has gone in to take a bath, and suddenly both my legs seize up in violent cramps, cramps in both upper thighs that turn my legs into these wicked, painful knots of constricted muscle. I have had this happen once before in my life—I was seventeen, and it was after a high-school football game that we had played on a very, very hot afternoon. This time the pain is almost unendurable, and I call out for Kathleen, who emerges naked and dripping wet from the bathroom to find me thrashing around on the floor, my legs out straight, eyes closed, jaw clenched; she pounds on my legs, tries to loosen them. And then I remember it was salt that we took during that football game, salt tablets for cramps, and I get her to go into the kitchen and make me up a glass of warm salt water. When I drink this, the cramps go away almost immediately. The floor of the living room is a chaos of plates, books, and magazines from where I have been thrashing around between the coffee table and the couch. Everything is wet from where Kathleen, ministering to me, has left water tracks from her interrupted bath. But I am all better. I stand up. I laugh. We both laugh about it. Kathleen tells me to take a couple of calcium pills, which are supposed to help muscles relax. Later she goes to pick up Shannon at the basketball game and I go into the TV room to watch the summary of the O.J. trial. Today the prosecution displayed the evidence of the bloody glove. I am in bed and asleep by 9:30.

February 20

No MONDAY blues today as things start off. The weather is unbelievably beautiful, almost balmy, and by mid-morning the temperature has climbed into the low seventies, where it stays for the rest of the day. The golfers are out. In fact, at one point in the early afternoon we look up to see one of them picking his ball up off the diningroom floor. It was so noisy none of us heard the ball come ricocheting in. The day has a confusing, chaotic aspect to it because of the addition of subcontractors; the roofers have started laying two-by-twos over our black paper in preparation for the steel "golf-ball proof" tiles that will be installed as the finished roof. And the plumbers are here too, the same father and son who laid out the preliminary work while we were setting the floor joists in place in December. The roofers are a father/son act as well, although the son is the boss of this duo; the father, a stooped man well into his seventies, climbs up and down onto the roof like an old pro, but says he's just along to help out—he spent his working life "monkey-wrenchin' around" with various trucking outfits. Other subcontractors and delivery people come and go, adding to the confusion. All day, Vern seems to be off in one corner of the house or another, with a set of plans out and a pencil, discussing the toilet drain in the guest bathroom or the return air duct for the heating system or the water bib for the ice-maker for the refrigerator.

Brian and I spend the morning moving a whole wall in the downstairs southeast end of the house. It has to be moved three and a half inches to accommodate a drain pipe. Apparently, it's less costly to move the wall than the pipe. It takes us about four and a half hours. But the work goes well. I think of all the remodeling I did in Southern California. This task has the same sort of fussy, imperfect aspect to it.

At lunch with all of us together, we have three father/son combos if you include Vern and Bud, and four if you think of me as Brian's surrogate Dad. We gather round to look at Vern's hunting scrapbook, which he has brought for "story hour," as he says. Apparently to put an end to all the who-shot-the-biggest-buck tall-tales, he now has photographic proof going way back into the early days. The very first picture in the album is one of his Dad, B.J., at about age thirty, a faded black-and-white eight-by-ten of the young frontiersman, with cowboy hat, boots, and old Springfield, looking like an extra from *The Birth of a Nation*. We are all gathered around in the warm sun. The lunch half-hour inches toward forty-five minutes as we look at the years go by, the dead bucks, the elk, the forked horned, the four-point, the thirty-six-inch spread, the hunters in their Pendelton shirts with guns and smiles, the decapitated heads of their deer or elk arrayed in front of them. And we see Vern as a young man, hunting with his father, standing beside the carcass of his first deer. And later, an older young man in his twenties, back from Vietnam, looking more like Christopher Walkin than Don Knotts, another day, another dead deer. Bear Mountain, Desolation, Cross Creek—the pictures are neatly pasted into the album and each one has been labeled and identified with a careful hand. "Don't say anything about my nose," says the caption under one snapshot, referring to an elk Vern shot, whose nose was bent around into a ridiculous position. In another section, we see a series of flash pictures of Bud and Vern and old B.J., in his late sixties now, all in funny hats and glasses and noses, celebrating Bud's twenty-first birthday. The story hour comes to an end. Brian has hardly said a word.

When we go back to work in the afternoon, hanging plywood on the gables, Bad Brian makes a reappearance—slow, teasing, passive, stubborn, and sloppy; maybe he was lured into his dark corner by jealousy and the fact that he had no hunting scrapbooks of his own to share. We muddle ahead.

February 21

Another day of balmy temperatures with just the slightest hint of rain in the A.M., although the sun is obscured through most of the day by heavy cloud cover. Spring is here. On the drive to work, I pass crocuses, daffodils, and most impressive of all, whole plum trees that are starting to turn pink with early blossoms. This is a local phenomenon, this pink plum-tree display that happens as we turn the corner after Valentine's Day. I look forward again to eight hours of working in comfort.

Vern said yesterday that nice weather "softens you up," and he was only half joking. Thinking about the photo album of all the dead deer and elk, I remember the flash I had during one of the first days of work with this crew back in September, the flash of hard-edged, rough maleness, of hands covered with calluses and old scars, of faces lined with care and beaten to leather by the endless cycles of rain and cold and sun and heat, of stooped shoulders and thin, sinewy arms, and beyond all of that a distant, detached, slightly grim outlook on the world—a feeling of having been molded down into a hard, resilient alloy. I see that in Vern's face still. In a way it's the same thought that makes my heart go out to Brian, if for nothing else, then for his young age, for all the tough years ahead, for all those days on the site that will slowly, inevitably, turn him into this same brooding character who will say to some other worker on some sweet, warm, early-spring day of the future, that pleasant as the air may be, it is to be held in suspicion, distrusted, because it can soften you up, sap your strength, make it that much harder when you come back the next time in the wind and rain and spend all eight hours cold and wet and miserable, doggedly moving forward, laying up timbers, chalking wet lines, lifting sheets of plywood, and feeling the rivulets of cold water flow down the insides of your sleeves as you raise your arms.

I am an alien in this world, playing at being this construction cowboy, outside myself, looking back in, trying to wear the right clothes, to blend into the corps of workers who again today number eight with the two roofers back and the two plumbers. Funny, I realize that when I graduated from Harvard I had never even heard of a Phillips head screwdriver. The first thing I ever built, the bed in the back of my hippie VW bus, came out all wrong because I never knew that a two-by-four was not two true inches by four true inches, but a half-inch shy on both sides. Who would have told me these things? My father was a Republican businessman, a Harvard-club man, and a golfer with polio damage in his right arm, whose only tools were the weeder and the trowel for the garden. My mother, an artist, was a kind of rubber-band, Scotch-tape, and clothespin improviser. The work, such as it was, that was done to our house, was done by rough-looking working men, plumbers and carpenters and electricians who came to the house in their old trucks, and seemed to me as alien as if they had just arrived from Outer Mongolia. Now I had become one of them.

Brian and I work all day on the gables. The Bad Brian showed up again this morning. I woke up in the middle of the night last night obsessing again about my difficulties with Brian and then this morning had time enough to meditate and turn it all over to my "Higher Power," as we say in AA, so that I arrive at work in a semi-positive mood and hold onto it through most of the day. To simplify things, I make myself his slave. And I wonder about that. Am I too passive, should I assert myself? There doesn't seem any way around it, and so he bulls his way ahead, making mistakes, fucking up, covering up, blaming me, ordering me around. I try to stay down within the protective cover of my mantra, I say prayers, I remind myself it's a job, just a job, and I take the drudgery with gratitude, I really do, and work alone at it, smelling at one point the fragrance of fresh-cut grass as the gang mowers move down the golf-course fairway, so thankful that we are not in Minnesota or New England, but that we are here in this green valley, and that the plum trees are already in bloom.

I spend what seems like hours hand-nailing off the plywood up under the eaves where the nail-gun won't fit and where I have to angle the nails in such a way that I can just barely get a bite into the top of the truss rafter, and still avoid dinging the edge of the fascia board, which is only nine inches away. I'm using my 23-ounce hammer, choking up on it as much as possible, sometimes cramped up on the edge of a ladder or the roof. The roof has become so much more reliable now that the roofers have added two-by-twos every foot or so; it's like having multiple toe boards. Even up on the high, small roof above the diningroom, the roof that caused me so much anxiety last week, I feel completely at home this afternoon, pulling large sheets of plywood up from the edge, nailing, leaning out, a little looser and easier. It's amazing what a difference an inch and a half of rough fir makes to my self-confidence in these high places.

Brian and I seem to fluctuate in and out of harmony. He has me jumping to his tune a lot of the day, but I don't mind it, and love the old roofer who, overhearing one of Brian's imperious commands, comments down from the roof above, "Awful damn bossy, ain't he?" to which I say, "Well, he was the brother to two older sisters, so he's getting back now for all the bossing he went through." Brian doesn't seem amused by this, and I realize I shouldn't have said it. "Somebody's got to be in charge," he says, and I bite my tongue. All day long, I try to think before I say anything.

At the end of the day, we are almost done with all the exterior wall sheeting. On the drive home, it seems more plum trees have come into bloom just in the eight hours since my morning commute. And it's still only February.

February 22

A cool morning. The weather forecast promises sun but there is no sign of it until mid-afternoon. But no rain either, and we proceed through a quiet, easy day. I am sluggish and slow it seems, but I keep up a pace, cutting and fitting the last of the wall sheeting, nailing off areas that were somehow forgotten when we first laid up the boards. The last piece I cut is a piece that looks exactly like a little house with a peaked roof and a front door. It goes into the top of the gable at the east end of the big roof; the door of my piece will hold the top half of the vent, a metal grating about one foot by two feet. The roofers are with us again all day, but no sign of the plumbers. The tin bender shows up in the afternoon and fashions scuppers and flashing for the patio and the little, flat roof above the studio room. The house is indeed starting to look like a house.

Vern and Bud finish up the staircase and then start fitting the windows into place, while Brian and I stay ahead of them, wrapping any remaining window frames in Tyvek. But we can't wrap the whole house yet because the inspector still has to come and sign off the nailing, to be sure that the walls have been properly secured, particularly the sheer walls, which are structural walls with huge, metal hold-down straps that are anchored into the concrete stem walls. The house has five or six of these sheer walls, where the nailing schedule calls for a nail every two inches around the edges and every six inches in the field.

I have no troubles at all working with Brian today. Only once really, after lunch, when Bud tells me about a place that needs nailing and another place where a small piece of plywood is missing. Brian says, "Do you want me to do that for you?" and I say, "Why don't you do it for yourself," a way of rebuking him that I immediately regret. He seems hurt and says, "Well, he told *you* to do it," and I say something to appease him and to keep the harmony going.

I am vaguely depressed today and glance at my watch obsessively. 10:20, 11:05, 13:42—I keep my digital watch on military time and fourteen minutes fast, so that the most rapturous moment of the day, for sure, reads 16:44. Still, the day moves along and in the end I have a fulfilled sense that we have gotten somewhere. I am just locking up all the ladders with the chain when Vern points out that the tin bender has been stranded on the little roof above the studio. "We'll be back tomorrow," I yell up to him, which seems funny at the time.

During the last half-hour of work, Brian and I exchange jokes. I tell him the one about the Mulligan, the one about the corpse with the cork in his ass, and the one about the Limburger omelet. He tells me the one about the guy who pisses in the beer glass, and the one about Jesus and God playing golf. We leave on friendly terms.

February 23

I played hockey last night late—a city league game, we lost 5–4 in overtime. Didn't get off the ice until about 12:15, and just before I did get off, I fell and took a wicked blow to my right side, particularly to my right arm, which felt wrenched out of shape, similar to the way my left arm felt at the end of January. When I came home, I was more wired than exhausted, and stayed up for another hour eating ice cream and watching TV before I finally got to sleep about 1:30. I woke up early this morning after only a few hours of sleep and now I feel like a zombie—I can hardly lift my right arm above shoulder level.

I get to work a little after 8:00 and Vern has already rolled out most of the cords. Brian shows up and Vern puts us on fire block detail. We laugh about fire blocks, because they seem in some ways the most hellish of construction jobs. They have an almost Sisyphisian aspect—just when you think you've put the last fire blocks in, a whole new area emerges. The blocks themselves are simple enough, just single flat blocks between studs, placed into walls between adjoining rooms with different ceiling heights. Their purpose is to prevent fire from spreading too easily between rooms or from one room up into an attic area and hence to the roof.

About 9:30, we get a break from fire blocking and go out to the front of the job site to load Vern's father's pickup with all the old throw-away wood in the scrap pile. B.J., the original patriarch, the same young frontiersman from the front page of the family hunting album, has grown old and stooped with a craggy, leathery face and a tentative, slightly bewildered expression. He doesn't say a word to us, and even though I would love to hear his stories of the old days, I don't talk to him. We're the workers, after all. He's here to see his son and his grandson and see how the new house is going. Vern once told me that B.J. in his prime had been the best roof-cutter in the business.

The pickup holds most of the scrap wood, and the process, which takes over half an hour, leads Brian and me back into the archaeology of the job, back through the recent scraps of plywood sheeting, through the butt ends of the rafters, the ceiling joists and floor joists, the TJI Silent Floor joists, then the darker tails of the pond-dried two-by-sixes we used for the walls, until finally we come to the concrete-coated stakes we used to kick off the stem wall and footing forms. At the bottom of the pile, we find a few of the original survey and lay-out stakes—all of it like ancient history now as we look up at the house, most of its windows in place, Tyvek on many of the walls, the roof ready for the final roofing tiles.

The building inspector shows up in the afternoon to look over the nailing so we can finish Tyveking the outside, but he throws up obstacles and objections, and while he is at it, a few more runs of fire blocks. A round-faced, middle-aged Native American with a baseball hat over his shoulder-length grey hair and a couple of animal bone necklaces, the inspector has a wristwatch band at least three inches wide done in ornate beadwork and a cellular phone that he talks into from time to time. Vern always refers to him as the "chief" inspector. He is making the rounds with Vern and Bud, pointing out all the extra work that has to be done. When he comes to the room where we are working, he just points to a spot where the sloping roofline outside intersects the middle of a bedroom wall inside and makes a "fst . . . fst" sound with his mouth while he sweeps his hand in the general direction of the slope. More fireblocks he means. Sign language. "I got enough to keep you boys busy for a while," he says, laughing. And we laugh too, but you can see that Vern isn't laughing as the inspector moves through the rest of the house, itemizing all the extra blocking, nailing and clip, bracket and hanger hardware that will now have to be added, much of it in difficult and hard-to-get-at locations.

By the end of the day, I am blocking with thirty-degree angles up and down the staircase and cramped over so bad I can barely stand up. I feel like some old greyhound that can barely make it out of his stall anymore, let alone race. I remember when I was younger, when I was in my twenties and early thirties, I could have danced around on a day like this, I could have been flying, feeling looser and easier and quicker as the day went by. And at the end of the day, I would stop for a six-pack of Rainier Ale in the tall, green cans (the "green death") and fire up a big, fat joint and party all night long.

Today at the end of the day, I can barely drag my sorry ass to my car, throw off my belt and suspenders, crump down into the driver's seat, and with a groggy visage and bobbing head make my way back across town to home. Shannon and Kathleen have gone off to Portland for the night. I take a shower and climb into bed and sleep for an hour. Nothing feels so good as the soft, foam mattress, the sheets, and the glowing warmth of the down comforter.

February 24

Friday again. I start the day off in the dentist's chair, getting the final work done on my crown, and show up late for work. I've notified Vern, but Brian thinks I just overslept, and I leave it at that for a while, to see if he believes it, to see if the idea delights him (which it does), and to see, too, if the idea is consistent with his idea of me (which it is). When he finds out later that I was only at the dentist's he seems vaguely disappointed. As if he hopes for more irregular behavior from me.

My body is still a mass of tight and twisted knots. I go right back under the stairs where I left off and spend the rest of the morning finishing up all the blocking and corner drywall nailing that needs to be set into place before we can get final framing inspection. Every part of me seems sore and ill-used. I'm like a worn-down, rubbed-down eraser at the end of a stubby little pencil. The morning, too, seems to drag by, and again I find myself checking my watch at odd times like 10:52 and 11:18, which seem the most particularly difficult times of the morning. It's always something of a miracle that noon comes at all. Fridays the clock creeps at a slower pace than usual.

At the end of the morning, I catch up with Brian again and we work together blocking off and then nailing two more sheer walls in the downstairs. Vern is nearby, sweating blood over the multifaceted and architecturally complicated front-door wall, the last wall in the house that remains to be built; one, it is clear, that he has been putting off and putting off until now, when there is no longer any way to avoid it. A head-scratcher for sure. The bottom plate has about eight different angled cuts as it snakes its way along the front of the house, but the top plate, with the same exact moves, must be projected onto the sloping ceiling line, making every stud, when they are finally cut, a different compound angle. Vern is over there cursing, sputtering, muttering, doing his Donald Duck routine, pissed off that boards are not meeting up like they're supposed to. He's wearing his old angle bevel down to a toothpick. Still, the wall progresses. We watch from a distance and don't say much to Vern—lay low, try not to bother him.

After lunch, he comes up to me and says, "Are you doing anything right now?" which is a ridiculous question—I mean, he's the boss, I'll damn well be doing whatever he tells me to do. But still he asks it in this incredibly polite way, as if maybe I was at home by the fire reading a book and he hated to interrupt me. When I say no, nothing in particular (I'm still on fire blocks), he assigns me to the job of working with Bud, placing the two windows into the two small front dormers, and then hanging the four remaining fascia boards off of the dormer roofs. This is an unprecedented disposition of labor. I feel the way I used to feel in my junior year in high-school football, when I wasn't a starter, but the coach would call on me sometimes, send me into the game if we were ahead by a lot or behind by a lot (I wonder which it is in this case), and expect me to perform up to my potential. The danger of fucking up, always lurking, seems heightened for me by this new arrangement. As in—here's your big chance, hot shot, don't drop the ball.

Brian seems a bit at a loss, and asks me, when I am already up on the roof measuring for the fascia boards, if I would like him to "finish off the nailing of the fire blocks for me." As if it would be some sort of favor to me personally to do the job, as if we weren't working on it together, and even more absurd, as if every job wasn't ultimately for Beverly, the client, and for Vern, the contractor. I tell Brian at some point today that I do not care what jobs I do, that it is all process, that it is all a question of working—or staying busy—for eight hours and then going home. Brian says that there are some jobs he likes better than others, jobs that are more demanding, more glamorous in some way, more technical, requiring more skill. I know what he means, I even agree with him, but I argue with him anyway.

And here I am being given a promotion, being sent up to the roof to work on fascia, to set windows, jobs that are Vern and Bud jobs. I'm being tested for more responsibility, and I had better shine like a star. I make the cuts pretty well, but do have a couple of nailing miscues that Bud monitors as if they were to be expected from the new guy. I think he has a picture of me as the chronic klutz and so, in that way that so often happens, I am almost destined to screw up if I am being watched by him. Still, we get along well up there for the few hours we're working together. And I bite my tongue and refrain from telling him that the last time I hung windows was with Jimmy Carter on a Habitat for Humanity project down in San Diego.

This happened when I was working as a segment producer for a low-end, short-lived PBS TV show called *The Nineties*. We heard that Jimmy was coming to the Southland, so we drove down there to meet him (this must have been around 1990, 1991) and to shoot tape with me as the host/carpenter. The price I paid for that three-minute run of tape was ten hours of hard, nail-pounding, Skilsaw-buzzing work in ninety-eight-degree heat, with a mob of gung ho Christian builders, most of them such novices that they were just learning that the sharp end of the nail is the end that goes into the wood. Jimmy was a different story. Possibly the original anal-retentive carpenter, he was using a slick, digital two-foot level and a chrome hammer with a curved, mahogany handle. Neat as a pin in his tight-weave cowboy shirt and H for H hat, he worked from inside, I from the outside, while we set this big sliding window. When he saw that the flashing was missing, he summoned Rosalyn over with a slightly peeved, imperial tone. She appeared in a minute, flashing and staple gun in hand, floppy sunhat back on her head, put the material in place, smiled, and hustled away to her next job. And later, when I dinged the window a bit with my big, waffle-face framing hammer, he frowned and handed me the sleek Chief Executive chrome model to finish the job.

But I didn't tell any of this to Bud. Already I have a reputation for making up giant, imaginary tales, even though I do only tell them true things that have happened to me. And really, I don't tell all that much; I keep my personal life as separate as possible. Still, my life, what little of it seeps out, so much of it devoted to drugs and art, so much of it spent in Los Angeles and San Francisco, must seem almost unimaginably exotic to a young, provincial man like Bud. So this afternoon, I am just a hammer.

The day starts to wind toward quitting time. A levity seems to infect everyone. The roofers are jolly, the heating and air conditioning guys are walking around smoking cigarettes and joking. We move a giant stack of lumber to make room for the crane that will come Monday to load the steel tiles onto the roof. While we are moving the lumber, Vern accidentally bangs my shin with a long two-by-four. It doesn't hurt at all, it's rather that Three Stooges kind of thing that happens with long timbers, sometimes, where the impact is not in the shin but on the head. But Vern laughs, an affectionate, easy, one-of-the-guys kind of laugh that is so uncharacteristic it makes me stop and stare. Only late on a Friday afternoon could he laugh like that. We depart for our weekends with light hearts and a fragile sense of bondedness, as if we have all survived another long week in the trenches and now it's off to Paris for a little R&R.

There is no sweeter time of the week than the drive home on Friday afternoon. I stop at the 7-11 and buy myself a large cup of coffee and a "Monster" chocolate-chip cookie and savor it all. The plum trees are bursting with color. The kids from the university are wearing shorts, the joggers along the running trail are intense and serious, and three or four trucks for sale are parked on the street opposite Dari-Mart. I study and enjoy every detail. When I get home, I take a shower and climb into bed for a half-hour nap before basketball practice with Shannon's hopelessly inept team. I love them all, and I love them the more after every loss. Just one win, one win for the season—that's all I ask.

February 27

MONDAY. A beautiful, warm and, in the afternoon, sunny spring day. My body feels somewhat recovered from the weekend off and I find I can move around pretty much pain-free, though I take a Nuprin in the car on the way over as a preventative measure. Now it seems I'm always hearing about people like Warren Christopher who come down with bleeding ulcers from taking "anti-inflammatory" medication, so I'm checking my stool for blood traces. Not really. I'm kidding! I'm just kidding!

Brian called me last night to ask my advice about changing jobs. In spite of the animosity and friction I feel, he truly seems to like me, and now, at least, to regard me as something of a mentor. The Good Brian has been back for the past few days. Seems a person in his church has informed him about a company that's looking for a framer. He'll start off at a dollar an hour more, $9 an hour, and has a good chance to go right up to $10 or $11. I was touched that he called me up for my advice, and told him that unless he wanted to keep working with me for sentimental reasons, he should certainly take the better opportunity. Secretly, of course, I was ecstatic to think he might be leaving. I had immediate fantasies of being off in some remote corner of the house, working by myself, hassle-free hour after hassle-free hour. We talked about how it was hard to read Vern, hard to know what he was thinking, hard to know what the future held for Marshall Construction. Brian and I joked around about all the things we have observed and he told me "at least I'll take Tommy with me," meaning Tommy Tyvek, our unofficial mascot.

When I get to work today, I'm the first one. Brian pulls in and then Bud, and we stand around for a bit talking about house fires before Vern shows. Lots of action. Brian has no time for his "little talk" with Vern about quitting. At one point in the morning, I count eleven people working on the job site. The four of us, two HVAC installers, the father/son roofing crew, and then three other roofers who show up with a stink pot to hot-mop the patio and the small, flat roof above the staircase.

Beverly comes and goes, a diminutive, middle-aged Chinese woman in skirt and blouse with black high-heeled shoes and glasses, making her way around the job site with what looks like steely concentration. She is regularly pumping change orders into the plans and these are always greeted with contempt, not sexist or racist exactly, but more client-oriented—the client first and then the architect are always the two main enemies of the carpenter. If the client happens to be an Asian woman, then the rift widens. These all-powerful outside forces, with a twist of the head or a flourish of the pen, can cause sometimes days' worth of work to be undone and redone in a different way. My attitude is really— so what? As I said to Brian last week, it's all process. As long as I'm busy for eight hours, I don't much care whether it's forward construction or reverse construction. I've worked for design/builders for years and they're *always* changing their minds.

Brian is in a super Good Brian mood today and we go over a lot of our mutual nonsense, the names of the ladders, the time I hit him in the head with a nail from the nail-gun, the times I've had my finger pinched when he was too eager moving some big timber, the time the Sawzall fell on my head, the pencils we have with their own personalities . . . on and on. I feel Bud's dark, lurking form off in a corner observing us and think of that line from *It's a Wonderful Life* when George Bailey goes into the tavern with his new friend Clarence, the angel, and the bartender says, "We don't want no characters around here." On a day like this Brian and I are both characters.

By the end of the day, Brian has still not talked to Vern about his plans to leave the job. He is hoping to stay on a little after work and talk to him when things are less chaotic. But as we're wrapping up the tools, Vern starts looking around for the stamped set of house plans and can't find them anywhere. The stamped set are the plans that have been approved by the city, stamped, notarized, and checked periodically by the building inspector. When you call for inspections, these are the plans you have to show to the inspector, and now, inexplicably, they are nowhere to be found. Vern is fuming. We fan out into different corners of the building and site, searching. Bud even asks me, in a pointed way, if I have seen them, as if I had somehow been looking at them and had taken them off somewhere. No sign of them. Now Vern is going off into one of his Donald Duck temper tantrums, and then Bud emerges from the Sani-pot with the missing plans held gingerly in his fingertips, the pages dripping with shit and piss. Someone threw the stamped set into the crapper and actually, he admits, he had noticed them when he used it earlier in the day. At the time he first saw them, he thought he was looking at a scrap piece of Tyvek down there. But now, when he goes back, he finds that what he was looking at was indeed the stamped set. Vern goes into a smoldering, ashen-faced rage.

Brian comes up to me with a lowered voice and says, "I'm still gonna tell him (about quitting the job). I like living on the edge." "Lotsa luck," I say, and I pack up my tools and head home before any more shit hits the plans, which are laid out to dry on the radial arm saw table and reek, even from thirty feet away.

I'll miss Brian, I realize, I'll miss him a lot in spite of all the petty annoyance I have felt over these months. We've managed to stay loose (at least most of the time) on a job site that may be the most uptight place I've ever worked. And at moments of crisis we are always so starkly revealed to be the labor force, cut off from, and in opposition to, the father/son management team. We are both privately glowing with delight about the plans ending up in the dumper. The madder Vern gets, the harder we have to work to keep from bursting out into fits of giggles. I can't look at Brian—he can't look at me. When he quits, I'll become a labor force of one.

February 28

Another clear, sunny day, although it starts out cold. Brian and I begin work on the back side of the house, in the shadows, putting up Tyvek, and my fingers are soon tingling with pain, even though I'm wearing two sets of gloves. As we come around the corner of the garage, into the morning sun, the temperature seems to rise fifteen degrees. This morning as we start, the temperature must be down at least into the high twenties. But by noon, the bright, warm sun has returned us to that wonderful sense of springtime everywhere.

Brian told Vern last night that he was going to leave and take the other job. Vern's only comment, apparently, was that he should "do what he had to do." Brian says Vern also asked at the very end how much the other company was willing to pay, and Brian lied slightly and said $9.50 an hour instead of $9.00, on the off-chance that Vern would try to match it.

Brian also tells me that Bud told him, back in September when we both started working for Vern, that he made only $9.00 an hour, which shocks me when I hear it now. I somehow assumed Bud was making around $12.00. I certainly assumed he was making more than I am, and now I see another reason for his surly, passive manner. He must hate this job on some level, hate working for his Dad, hate the feeling that he isn't going anywhere but down the same, long, dark tunnel his father, his grandfather, and his great-grandfather traveled before him—the tunnel of construction misery, of ill-fitting boards, of untrustworthy levels, of ungrateful clients, unreliable subcontractors, uncooperative inspectors, and stamped sets of plans that end up in the shitter.

At coffee-break, Bud goes into a long tirade about his best friend's boss, who makes his best friend work ninety hours a week on salary, who won't buy new equipment, who has workers who are quitting, who's a penny-pinching skinflint. After break, Brian and I wonder—was he really talking about Vern and he didn't even realize it?

We're off in our own world with the Tyvek and we start to develop the true story of the sabotaged plans. Bud, whom Brian has always hated, now becomes the butt of the worst jokes, another classic case of the boss's son taking all the heat from the workers who share the common bond of a common enemy. Bud does ride me, in a way, but he has never gotten too deeply under my skin—not the way he has worked his way under Brian's. And Brian and I, with only a few more days of working together, are now trying to pile up a last slew of common jokes and amusements. And so the story of the plans, as we work it out, is that Vern gave Bud the stamped set of plans to study and Bud took them off to the shitter, "his office" as Vern has sometimes called it, and that after he took a shit he became confused, wiped his ass with the plans and returned a neatly rolled length of toilet paper to the plan box. Also, we consider the moment when the inspector returns: "Are you sure you want to see the *stamped* set?"

We work hard, we make good progress, we Tyvek almost all of the house, Brian replaces a faulty sheet of plywood, we rip cedar for the edge banding, and we try to review the great events of our working life, going back to the fall and to the Baxter house, the house where we started. Our list of great moments at the Baxter house includes: (1) the "Old Sparky" incident, when Brian was up on the roof cutting an oversized rafter in close quarters, in fact with the Skilsaw between his legs, and the switch caught on fire and sent flames shooting out, just inches from his balls; (2) "The Shop Vac Incident," when the client, Ray Baxter, showed up during a bone-drenching rainstorm and tried to get the water off the ground-floor plywood with a shop vac even as it was pouring down in literal sheets through the open roof; (3) "Big Red," Brian's red carpenter's pencil that lasted through the entire job; (4) "Beat the Dealer," an ad for the state lottery featuring an Andy Devine–type card dealer and a coquettish lottery player who always won, which got stuck in our heads while we were skip-sheeting the roof and which we repeated verbatim to each other for weeks; (5) "Number Nine," our way of measuring and laying out the one-by-fours on the roof; and (6) "The Time I Hugged Ray," the incident in which Ray was telling me how to do my job and I jumped down off my ladder to give him the big hug and tell him how much I appreciated his help. These are the standout moments, drawn from a whole list of memories. Brian says he's going to bring a notebook tomorrow so we can write some of them down. He also says that on Friday he's going to have his Mom bake a big cake to bring in to work, as a goodbye present to everyone.

The day goes well, and at the end of it, I still feel halfway decent. As the house progresses toward completion, the boards get smaller, the nails get smaller, the work gets more technical and minute, and the physical attrition of raw framing diminishes as the workload shifts more toward finish carpentry. My body needs the break.

March 1

Another day in the countdown to Brian's departure. I arrive a little late, and Vern is still not at the site. Brian is there in his truck and tells me, as he is putting on his work shirt and hat, that the boss of his new job may have an opening for me. At $12.00 an hour. Do I want it? I hedge a little. I want Brian to think I'm at least interested. Two dollars an hour is a significant sum. But I tell him my first instinct is loyalty toward Vern. Also, and I don't tell him this, I'm curious about how the dynamic of the job will change with Brian gone. With the father and only two sons.

All day, Brian makes bitter, dark-humored remarks about Bud, and I'm increasingly uncomfortable with it, afraid for one thing that Bud will overhear us. We work in all kinds of places during the course of the day and so does Bud. Inevitably, sometimes we will be working on one side of a wall and Bud will be working on the other. I try to smile in a neutral way when he says something about Bud. At one point I say, only half joking, "Brian, I think you have to work on your compassion." Bud has hardly ever really done anything to bother me, and I don't have the same animosity toward him at all.

We work ripping all the pieces of outside trim wood in the morning, and in the afternoon, while Brian paints them, Vern gives me another complicated job, this one more complicated than the fascia boards on the dormers, and with the added bonus that he gives it to me alone. It involves framing two forty-five-degree walls (in plan, not elevation) and then a closet built into the sloping roof. Added to the puzzle is the fact that the heating pipes run through the middle of it all and will have to be boxed out. There are angles in all directions. Still, it is nothing I can't handle, and as I get to work on it I realize it's the first time in a long time that my brain has been thoroughly engaged in a project. Time flies by and I feel I have to move fast, have to prove something. I know what I'm doing—but it takes time. I make a few minor screw-ups along the way.

Brian shows up after a while and cuts for me, and we have a pleasant afternoon, warm and easy, up in the guest bedroom and bath working on this fussy little framing project. We continue to talk about the different things we remember from our six months of working together. And I think Brian senses that I won't be moving over to the other job with him. Although I will tell him, when the time comes, that it may be an option for the future.

For now I'm committed, at least in my own mind, to finishing Beverly Chin's house. I'm curious to see how it will go from here on out. I assume we will be doing the siding. And then maybe the case and base, the interior doors, cabinets, who knows how the last touches will end up. The job will certainly be different without Brian around.

At the end of the day, he takes me over to his truck and shows me his high-school letter jacket, his prize possession, which looks like the kind of thing Pat Boone would wear if he were ever made a Russian general. Medals hanging all over it. Football and wrestling, state and division championships. Brian's mighty proud of the thing. Tomorrow is Thursday already. The week has flown by fast. It's been another day of warm, sunny spring weather. What a blessing.

March 2

Somehow today almost feels like Brian's last day, even though it isn't. Who knows—he may stay on through next Tuesday, he told me this morning. I am trying to come up with a going-away present for him and have all kinds of ideas, but so far nothing concrete.

We finish with the complicated framing details in the second-floor guest bedroom in the morning, and then switch over to siding. Bud is brought out onto the siding crew, and with the three of us working together, the complications become numerous. The friction between Bud and Brian is palpable. At one point, Brian pulls down what we call his Dari-Mart hat, a hat that becomes a ski mask and makes the wearer look like a holdup man. He has his hat down in this position for maybe half an hour, arguing with Bud at every turn. They nit-pick about whether a piece of siding is on the line, about how best to get the siding straight, about how much of a gap to leave at either end of the board. Most of their arguments fall in the one-sixteenth to one-thirty-second of an inch range, over a process that could probably absorb errors of up to a half-inch over twenty-five feet before the naked eye would notice anything askew—or "crookit," as we seem to say on this job.

When he first comes out to work with us, Bud says he's been "banished to the outside," and on the surface he seems to be talking about the fact that it's colder outside where the wind is brisk and a bit nippy (it never does warm up much today), but clearly what he means is that, like a sentence to the Gulag, he has been sent out to work with Brian and me. Mercifully, Brian has to go off to have a doctor look at his infected toe at 3:00, and Bud and I are left to work together for the rest of the afternoon, a prelude to what the working situation may be after Brian leaves.

"I don't know how you work with that guy," Bud says, when Brian has driven off. I tell him that it took me two months of sweat and anguish to figure it out. The key to working with Brian, I say, is that you never tell him how to do something. You let him take charge and go as far as he can, let him make all the mistakes you see clearly coming, let him re-do work if that is the consequence, give him the illusion that he knows how to do everything, and stand back, even in a totally subsidiary role if necessary. Ultimately, particularly on production work, the pace will catch up and even surpass what it would have been if you had bullied him into a passive-aggressive rage at the outset by insisting on doing it right. This is how I spent those first two difficult months, I tell Bud, learning that by this sometimes-humbling leap of faith, the job would eventually stumble and then succeed. Bud shakes his head—the trick is almost too complicated to explain. Still, we both feel relief at the thought of Brian's departure. "The job'll be a lot quieter—that's for sure," Bud says.

And later Vern comes by to visit, hanging out in a very un-characteristic way, his arms folded, looking at our siding work, chatty, disclosing more about things in general than I have ever heard him disclose. He says he has another remodel job ready to go, that he is sorry to lose Brian as he had hoped to split the crew into two, with two (which two I don't know—me and Brian, me and Bud, Bud and Brian?) going to the remodel and two staying to complete what will mostly be finish carpentry at Beverly Chin's from here on out. Now it seems likely that a new person will be hired. The prospect of a new person coming onto the crew disturbs me. Anyone may arrive. A new man, a new person to break in. Who will the universe pick this time to work for Marshall Construction?

Tomorrow may be Brian's last day. He says he's going to bring a giant cake with a fountain coming out of the middle and play his Ray Stevens CD on the boom-box for a lunch-time farewell celebration.

March 3

Another Friday, another week, this one a lot more manageable than the last; the work is lighter, the pieces are smaller, the physical effort is relaxed. Not the mental effort, though.

Bud, Brian, and I are together all day on the back side of the house, laying up siding, Bud working solo and Brian and I working in our usual tandem. This was going to be Brian's last day but he informs me when he arrives that he will be with us through next Tuesday, which gives me a little more time to put together my farewell package. Still, it's like we're in a lame-duck administration. And the presence of Bud has brought out the Bad Brian big time.

Everything comes to a sticky halt about halfway through the morning when we discover that one of the corner trim boards Vern put up is an inch and a half too low, and that we have a problem with leveling up the various sections of the wall we have now completed. Theoretically, you have to have the bottom board of siding at the same level all the way around the house. From that one board, everything goes up to whatever levels the house achieves as it moves from wall to wall. With the bottom board at the same level, all other boards line up in perfect parallel lines that meet each other everywhere they need to meet each other. But with one of the corners set too low, everything's fucked up. And my first thought is, thank God it ain't my fuckin' fault.

Vern is summoned. There are at least four different ways to solve the problem and a discussion ensues, which I withdraw from as fast as I can, leaving only three. When Vern leaves to go get the level, Brian and Bud get into one of their snits and the tension ratchets up another couple of teeth.

To make matters worse, Brian, apparently as a special deal due to his impending departure, has been given permission to play his new boom-box while we work, and Ray Stevens is back there at an ear-aching volume singing a song about how we're all driving Hondas and Toyotas and Kawasakis and working for the Japanese. Ray Stevens is kind of the Weird Al Yankovich of country music. He's enough to make anyone crack and you can see Vern is near the breaking point.

The leveling problem, however, is now shifted over to Bud's side of the wall, where it will be dispersed in small increments over ten or fifteen siding pieces. Brian and I go back to our side. Slow and goofy, Brian is trying to get the jokes going. Maybe it's me. I can't be myself around Bud either. I've been thinking a lot about how I finally figured out how to work with Brian, and I realize it is complicated and involves a combination of patience, humility, reverse psychology, and a capacity to be even more passive than Brian. Kathleen gave me the key when she told me, "Easy does it." But now, as the end of my six-month term of working with Brian seems to be in sight, I have these fearful fantasies that he will show up one of these mornings and tell me his new job has fallen through and that he's still going to be working with me. In a way, I've done it so well I've ended up convincing myself that it's fine, that it's actually okay to work with this guy. And really, when we're on the right kind of job and Brian is being the Good Brian, it is fine. He works like crazy when he wants to, we cover a lot of ground, he's funny and charming in his own narcissistic way, and some days I come home from work feeling like I had a pretty damn good day.

But on days like today, when the Bad Brian is in the ascendant, it's a torture to get through it all. We hack and nail and snap-line our way up the wall, piece by slow and brutal piece. It's like the allied armies working their way up the Italian peninsula. We never seem to hit a rhythm with the work. And always, Bud is off in the wings, ready to swoop in and comment on any of the screw-ups, an exercise in futility where Brian is concerned, since the information will only slow him down further, make him even more intractable, steer his cuts off square, bend his nails, force him to tangle his chalk line into knots, play Ray Stevens louder, and generally make himself as inefficient and obnoxious as he can within the disguise of still doing an honest day's work.

Mercifully, I have to leave at 3:30 to make it to Shannon's

basketball game, which I am coaching. It's a tournament game, the last game of our season, and we lose it by three points, a real heartbreaker. In the late evening, around 9:30, I feel filled with a kind of piercing sadness, partly for my basketball kids, who have not won a single game all season, but maybe too for Brian and Bud and Vern and me and all of us, us humans with our petty jobs and squabbles, with all this low-level suffering that seems such an integral part of everyday life. No big excitement today, no really good laughs, just gray, synthetic particleboard siding climbing slowly up the walls of the house of another relatively rich person in America in the late twentieth century. Someday before I leave this job, I have to have a talk with Beverly Chin. I want to hear something about her days as a medical student in Beijing in the 1960s.

March 6

MONDAY, Brian's next-to-last day. The day starts off cold, with us working on the back side of the house. As I drive up, I'm struck by how far we've come. The roofers have almost finished the roof sections that face the street, which are now covered with a dark brown, fake tile, steel-composite, sectional roof—golf-ball proof, or so we've been told. The walls are all in place and the windows are all in place and the sliding doors are all in place. In the few gaps where you do not see Tyvek, you see the yellowish surface of the composite sheeting. The electricians are inside pulling wire through the walls. The plumbers have finished with their rough plumbing, all the copper water lines and the black plastic drain lines are in place, two tub-shower units and a Jacuzzi are also installed. Vern has been working for a few days on a massive fireplace surround that stands as a seventeen-foot-high island in the middle of the livingroom.

Outside, we're the three siders. Bud seems in a particularly good mood today—maybe it's the prospect of Brian's departure. He rattles the bottom of my ladder every time he passes underneath it, sometimes giving me a rush of adrenaline that spreads from the soles of my feet up to my forehead with a pulsating, warm and cold sweating, heart-quickening sensation.

I'm working up at the very top of the laundry-room wall, just below the small, flat roof that covers the staircase area. My head is about twenty feet off the ground, and for the most part, I'm pressed in tight against the ladder and the wall, reaching tentatively as far as I can in every direction to nail off the siding, which goes up slow, one piece at a time, each piece covering about six inches of vertical wall space and, in this case, just a little less than sixteen feet horizontally. Brian is on the other ladder. We snap a chalk line for each board, cut it short enough to leave a one-eighth-inch expansion gap at each end that will be filled with caulk, lay up the board, and then nail it from one end to the other to avoid the rippling effect as per the manufacturer's specs. The nails are eightpenny galvanized nails and they have to be sunk into studs. Stud finding becomes a major part of the process. If you're not hitting a stud, you don't know it until the nail is already over an inch deep, when, instead of biting into tough, weathered two-by-six, you get an "air ball," come out into empty wall space behind the sheeting, and have to pull the nail out and probe around till you find the missing stud. The studs are on sixteen-inch centers, so once you find one you usually find the rest, though this is not always the case. Sometimes we have to climb down off the ladders and go study the wall from the inside, where the framing is still exposed, to see where the hell the studs are.

Brian has a little energy in the beginning of the day and we move along at a crisp pace for a few hours, but by 11:00, he's running out of gas and everything becomes slow and tedious, with long monologues on movies he saw years ago, things he did last fall, things that happened to him when he was a kid. I stand on my ladder and listen and wait patiently for him to put another nail in the wood. You cannot tell Brian to hurry. When he gears down, he gears down, and any prodding just gears him down further, sometimes all the way down to the creeper gear. In this mood, everything becomes a little goofy to him.

One of the stages of our process is getting the end of the chalk line from one person to the other. We are on ladders and usually ten or twelve feet apart. Brian tries a number of different techniques, and for a while seems far more interested in the challenge of throwing the line than he is in the chalk line itself. Sometimes he takes out ten feet of string and throws the end of the line like limp fishing line at me on the other ladder; sometimes he holds the line and throws the chalk box at me like a bullet. At one point, he wraps the end of the string around his cat's paw and throws that. I drop most of his throws, so then I have to go down, pick up the end of the string, and climb back up again. I'm not going to miss this crap.

I know this: When I'm at work, I like to work, to get into the rhythm of it, the perfection of it, the uninterrupted flow of it— on a really good day, the mystical, mantra-sounding, deep spirituality of it. Brian in his current mood is a walking, nailing, ladder-climbing, standing-around-talking, constant interruption. Interruption becomes his primary *modus operandi*.

Bud stays away, out on the edges, caulking the seams. Vern appears from time to time, usually catching us standing around, and says his usual, "How's she goin', boys?" Because the walls are all up, he can no longer see us unless he comes looking. Tomorrow is Brian's last day and he claims he is going to bring a cake.

March 7

I spent last night getting together Brian's farewell presents and I wrap them this morning, before going off to work. There are five in all: (1) "The Silver Soldier," a silver pencil that I had for a long time through the first part of the Beverly Chin house. At one point, it broke and I repaired it with several wrappings of duct tape, a grey mass of squishy rubberized fiber that I chewed on for several weeks before losing what was left of it to the bottom of my nail bags. For Brian I put it into a gold jewelry box. (2) An exploding golf ball that I got out of my father's desk after he died. (3) Two cabinet knobs made from threaded pins and golf balls from the site. (4) A bag of sixteenpenny nails with a computerized label, an extension of a joke we have had about how Brian is hauling off nails at the end of each day in his nail pouches, and reselling them in small packages at the flea market. The label says BRIAN'S BAG O'NAILS. LOTS FOR CHEAP. YOU BUILD 'EM. THERE'S PLENTY MORE WHERE THESE CAME FROM. And (5) a wallet-sized card, laminated with the miniaturized logo of Tommy Tyvek and the words "Official Member. J.F.D.I. School of Construction. Tommy Tyvek (Pres.)" with the signature of Tommy Tyvek below. J.F.D.I. stands for Just Fucking Do It. I put the five presents into a True Value Hardware bag and head off to work.

Another beautiful day, though cold at the outset. We go right back to our siding project, and I feel as though I am counting the minutes until I will be free of Brian. He keeps making references to a surprise he has for me at lunch-time, but I think little about it. Maybe he has bought me a new twenty-five-foot tape, since my current tape is in such poor shape.

At lunch-time, Brian's wife, Jennifer, arrives with a cake she has baked for the occasion. Ever since Brian decided to leave, he has been making plans for this going-away cake. It is a beauty—a large, square, white cake with a golfer at one end of it and a small house at the other with a broken window. A real work of art! There is also the name Tommy Tyvek along one side and a picture of Tommy Tyvek on the top next to the golfer.

I bring out my presents, but now I start to hear something that I only gradually grasp. My resistance is so strong that at first I don't hear it, and then I do—Brian is not leaving after all! My worst nightmare is coming true right before my eyes. I can hardly believe it. Even though he has the cake and I have given him all these presents and he is opening all his presents right in front of me, he tells me that the truth of the matter is that he has decided to stay. It takes a while, in fact most of lunch-hour, for this to sink in. At first, I think he's just joking, "just kidding," as he so often does. And only after lunch, after Jennifer has gone, after we are back out on our wall putting up more siding, do I get the full story, that at the end of work yesterday Vern came up to Brian and said, "What do I have to do to get you to stay on," and Brian said, "Well, I guess you have to match the money I was going to be making at the other job," and so Vern agreed and has now given Brian a raise to $9.50 an hour.

I am plummeted into an almost catatonic depression. I feel as though I have been in prison for a long time, maybe in some kind of dark hole, and then for an instant I have been given a glimpse of the light, of what the light might look like, of what freedom it-self might look like. I have been entertaining these pictures of myself, working alone in a sober, steady, quiet construction world, and now all of a sudden that door has been slammed shut and here I am, back in the black hole with Brian, with Brian for-ever—it's some kind of horrible cosmic joke. I realize that my presents have kind of tipped the thing in a way. Such graphic evi-dence of my eagerness, my own celebration that I was about to be rid of him. I'm so happy he's going I've come up with all these crazy presents to give him. And now he's staying! It's sickening. We're back into the same old drudgery, the same old lackluster work, Brian's same old habits and patterns.

I suffer through the remainder of the day, trying to hide my disappointment as best I can, and when I get home and Kath-leen gets home, I tell her everything. She is adamant about the unfairness of it all, particularly that Brian got such an enormous raise, and she is telling me that I have to talk to Vern about it. And now I think that somehow I will have to, will have to at least ask him for a raise myself, something I absolutely hate to do, it strikes at my deepest feelings of fear and alienation toward authority and I want to just say—what? This is some kind of incredible, spiritual challenge. I am going to have to boost my practice, meditate twice a day, go to ninety AA meetings in ninety days, do whatever it takes to survive this. The last thing Kathleen said to me this morning as I was showing her all the presents I had made for Brian turns out to have been an ominous prediction: "You're putting so much energy into these presents," she said. "Now you're never going to get rid of him." And I said, aglow with the thought that I was soon to be set free, "Oh, he lives way up at the north end of the valley, I'll never see him again." Now, tomorrow, instead of having my first day with the new three-way dynamic, just me and Vern and Bud, I'll show up and it will be another day just like all the other days.

Rain is predicted for the rest of the week.

March 8

Today I am in the pit of despair. I get to work a little early and I have to use all my spiritual resources to stay centered and keep myself from literally driving away from the place. It's wet and cold with a forecast of rain through Friday. Vern pulls in about five minutes after me with all the corner pieces we've been waiting for so we can get on with the siding job. He tells me that Carol Anne is back in the hospital, that she stopped breathing twice yesterday. It does not sound good, and Vern has that same haunted look he had last November when she first went into the hospital after her tumor was discovered. Bud, of course, is with her and will not be in today.

Brian shows up fifteen minutes late, and after the preliminary work of setting up the tools, we go back out into the patio area and continue with the siding where we left off yesterday. Desolation clings to me like a cold wrap, and it must be obvious, particularly to Brian, as I am monosyllabic, even surly. He picks up on it pretty fast. And I feel myself wallowing in it. The whole scene seems depressing. The inside of the house is a chaos of sawdust, butt ends of boards, ripped stock for the outside, sixteen-foot-long stacks of siding that we're slowly pecking away at, scraps of wire everywhere where the electricians are leaving their trail, bits of roofing material, old crumpled-up pieces of Tyvek and black paper, nails, cardboard from the skylights and the other appliances and lighting fixtures that are slowly being set into the various ceilings of the house, big Mac wrappers, Diet Coke cans. We are cutting siding in the dining room with the huge fireplace structure beside us, looming like some forlorn wrecked cathedral ceiling—the place has a bombed-out look. Chaos, mess, disruption. The cold hovers inside the house long after the air outside warms. We work under the roof overhangs, and as it continues to rain, rivulets of water come down onto us in our positions on the ladders.

Brian is thankfully quiet and seems to respect my mood. He wonders (ever the intuitive) if it may have been triggered by his change of plans—but I tell him no, no, that it's just my regular despair cycle, that despair is part of the artistic process, that great creative ideas are spawned from such dark spirits. He tells me he hasn't been in a bad mood since his last year in high school.

I obsess about asking Vern for a raise, but I can't picture the scenario. Self-respect seems to demand it; low self-esteem makes it almost impossible. In my imagination, the best I've been able to come up with so far is to say to him casually at some point, maybe when he hands me my paycheck: "So, do you ever give anyone a raise around here?" Or maybe: "What's the company policy on raises, anyway?" Or maybe: "So, whaddya have to do around here to get a raise? Threaten to quit?" I hate asking for raises. I feel like he should just give it to me. And then, when he doesn't, I start to seethe with resentment. I can hardly look at him.

[137]

At lunch, I do something unprecedented—I tell Brian that I have to run some errands and I drive away a few miles to sit in the parking lot of a mall and eat my peanut butter and jelly sandwich and watch the rain stream down my windshield. The half hour of solitude is almost blissful. I sit in silence and eat. I do not even read. I try to meditate for a few minutes, try to get my spiritual house in order, count time, think to myself: I'm halfway through the day, halfway through the week.

After lunch, when Brian asks me where I've been, I tell him I went to see my lawyer. We crawl across the long, cold afternoon, putting up endless sheets of gray siding. All this forest product material. This stuff is stamped with a grain pattern to make it look like the real thing. The pieces lap over each other, leaving six inches exposed on each course and that's our pace as we progress slowly up the wall, a little over half a foot at a time.

I'm in my rain pants all day, even though it doesn't rain in the afternoon. I feel sullen and surly and miserable and can't wait to leave at 4:30. I keep thinking this: If Vern's gonna pay me ten bucks an hour then that's all the work I'm gonna give him, and so I drag my heels and take extra trips to the Sani-pot. Despair. In the sixties there was a famous R. Crumb comic book by that name. And a character named Pete the Plumber who finally couldn't take it anymore and flushed himself down the toilet.

March 9

High winds and rain predicted today, a weather system described as "another big storm off the Pacific." The thought of terrible weather gives me some comfort, as if the extra adversity of just moving around outside, wet and cold, will take some of the sting out of my humiliation, boredom, and passive despair.

As we are rolling out the tools, Vern says, "Here, I got a present for you," and flips me a round, sticky, heavy-duty piece of yellow plastic that's designed to go on your tape. "Carpenter's Scratch Pad," it's called. Both Vern and Bud have them and in fact I've been meaning to get one myself. You can write dimensions on it and then erase it with your spit. Brian and I use old scraps of two-by-four or plywood for our scratch pads now. Mired in my resentment, my reaction to this gift is secretly hostile: So, you give Brian $2.50 more an hour and you give me a $2.50 tool accessory. Outwardly I say, "Thanks a lot! I've been meaning to get me one of these."

Brian shows up half an hour late, having run out of gas on Spyglass Road. He says he didn't have the cash and was hoping to just make it to work. I drive him off to fill up his gas can, and while I'm driving I'm thinking, I hope Vern sees this, sees what kind of shit his golden boy, Mr. $2.50-An-Hour-Raise pulls all the time while his steady, reliable, loyal, show-up-to-work-on-time good old boy just keeps hacking away for his $10 an hour.

Late by almost an hour after all is settled, we start work on the siding for the front wall, a high-visibility wall just to the left of the front door. It's a wall about sixteen feet by thirteen feet with eight windows, four long, wide windows below and then four smaller, matching windows above. The windows will be the southern exposure windows into the diningroom, which is a long narrow room in the center of the house, dividing, in a sense, the living quarters from the working quarters. The rain is coming down in torrents and we are directly under a small, gutterless roof that is pouring a sheet of water down directly onto our heads. In addition, a scupper that drains water off the upstairs patio area is without a downspout, so it is providing an even bigger flood down the left-hand wall, directly in the corner where we are working. After some prodding from Vern, we set up a temporary shelter under a piece of clear plastic and work for a while in relative comfort.

Because we are right at the front door, we are observed by and are observing in turn everyone who comes and goes from the job. If anything, the parade of subcontractors has increased. Beverly stops by two or three times in the first few hours, consulting with Vern, consulting with the electricians about outdoor lighting placement, consulting with the painters who have showed up for a few minutes to prime the corner boards, and consulting with anyone else who happens to come by. Brian and I work away as best we can. The wall requires a lot of small, somewhat tricky pieces between all the windows, and we do not make dramatic progress. The rain comes and goes. When it goes, the sun pops out for a few minutes and cooks us under our clear plastic awning. All morning I'm putting clothes on, taking them off, putting them on again. The work is slightly demanding and gets my attention.

At lunch, Jennifer shows up and takes Brian off to a restaurant, while I sit in the front room with Vern alone and try to imagine myself asking him for a raise. Why is it that I hate to ask for raises? Is it because I grew up in a rich household, a household rich with old money and saturated with New England Puritanism, where it was considered distasteful ever to mention money at all? As topics of conversation, money and sex were equally taboo. Or is it because I have such a low and self-doubting sense of myself that I find it hard to imagine that anyone would think I was actually worth more money? Far easier for me to believe, in this case, that Vern would secretly like to give me a "lower." At any rate, I say nothing at all on the subject and go doggedly back to work at the end of my half-hour lunch of chips, a hummus sandwich, an apple, an orange, and a cup of black coffee. I hate myself for being such a coward. Ten times the topic of the raise is on the tip of my tongue and ten times I chicken out.

Brian shows up fifteen minutes late again and I think: See, Vern, this is how your golden boy treats you.

My neurosis is out of control. To me, Vern as boss is basically

God, and as God he is the stern and unforgiving and punishing
Old Testament God who knows all my secrets, all my fuck-ups,
all my failings and shortcomings, and worst of all, he has access
to my innermost thoughts, which, among other things, include a
big fat resentment against him. In my wildest, most needy fantasies, I want him to come up after work and say to me, "Joey, I
just want to tell you how great you've been doing, how happy
I am to have you on the crew and working for me, how much I
appreciate all the extra effort you've put in, how much I appreciate your willingness to work with Brian day in, day out, and for
all that I want to give you a $2.50 an hour raise because I want to
be sure that you'll be working for me for a good long time." And
I know that ain't gonna happen in a million years.

We finish with the front wall and move around to "the swamp,"
an evil-smelling area at the east end of the house where a large,
flat, windowless wall rises out of the thickest, blackest, wettest
mud on the site. Now the rain starts to come down in tubfulls
and we sit with our power cords under two to three inches of
water, the sawhorse legs up to their withers in mud. I don't know
why we're not electrocuting ourselves in these puddles.

For some reason, my mood brightens. Maybe it's because the
rain is so loud I can't hear Brian's constant chatter. Now, on a
long, flat wall, we make better progress and we already have as
much siding laid up in a few hours as we did after our long, slow
morning on the front.

At the end of the day, Vern calls work off ten minutes early
because of the rain and even that gesture toward kindness and
understanding of the workers' plight seems, if anything, an insult. "Ten minutes," Brian says, "I mean, why not just finish up
the day? What's ten minutes? We're already soaked through." My
tape is so messed up from the rain and the mud that it takes me
two or three passes and some serious pushing just to get it to re-
tract back into its case. At home, a hot shower is heaven.

March 10

Friday again. The last day of the longest week in history. I'm in a catatonic state of despair again this morning as we pick up where we left off yesterday, on the long, flat, gabled east wall.

Brian seems buoyant, and while he talks and talks I try to make an effort, through the first few hours, to say absolutely the bare minimum. I even analyze his technique, his M.O. How is it that he can actually keep his mouth moving all the time? One of the things he does is ask me questions. Things from his school work like: "Do you know why it is that competition is fueling growth in small markets?" Or "What's that song about the 'Phantom 409'?" Questions I almost have to answer. After a while, he picks up on the fact that I am in one of my "creative" moods again, and instead of trying to rouse me to conversation, he starts singing these Christian happy-talk songs he must have learned in kindergarten Sunday School, songs about blowing all the clouds away 'cause life's not so bad after all, songs that absolutely grate on my nerves more than anything. But, like that old guy said to me so long ago—it's not the work that gets you, it's the people.

After a few hours of this, the rhythm of the work starts to pick up and in that I find some comfort. We are still in the rain, and no matter how much plywood and cardboard we throw down into the mud at our feet, it still seems to ooze back. As we get higher on the wall, the ladders are brought out, and their legs, pinpointed with weight, sink like sharpened stakes down into the soft ground. Every time I move a ladder, I have to strain and fight to lift the legs back out of the grasp of the mud. And when I set it down again the two legs sink at uneven angles. I'm using "Shaky Jake" at first, but he is so shaky with the off-angles that I give him up and go over to the "No-name" ladder, a thin, light-weight, slightly cranky extension ladder that serves me better.

At lunch, I sit silently with Brian and Vern. Bud always leaves for lunch and I wonder where he goes. Does he do what I did a few days ago—just drive to the nearest parking lot to eat alone so he can eat in peace? It's hardly peaceful where we are. Brian goes on incessantly with stories of things that happened to him in high school, a long story about giving a friend's father a red-hot fireball candy, another long story about a kindly Dari-Mart lady who used to give him free corn dogs when he was down on his luck.

At the end of the lunch-hour, I see Jenny Blair in the front door talking to Vern. What is this wild coincidence? Jenny Blair is my landlady from the first place we lived when we moved up here three years ago. I've been carrying around an ugly resentment toward her and her husband, Kevin, ever since we moved out of their apartment and they stiffed me for the cleaning deposit, charged me exactly $200 for miscellaneous clean-up details they claimed I had overlooked. I worked my butt off to clean that place—I had heard they were sticklers and I must have spent sixteen or eighteen hours scouring, scrubbing, mopping, and polishing. And now here she is talking to Vern. It's definitely her. I thought I spotted her once before about a week ago, but dismissed it as an aberration. Her hair was different, and I was thinking, maybe it's not Jenny Blair, maybe it's a close relative, a clone, maybe a coincidental doppelganger of some kind. But this time she is unmistakably herself. She is on the other side of the fireplace about thirty to forty feet away, talking to Vern, and handing him a roll of what look like plans and permits. And then it strikes me—is she a client of his? Is this the new client, the new remodel we're headed toward? This would be too, too weird if it were true.

With about five minutes left in lunch-hour, she leaves and Vern comes back again and sits by the fireplace, unrolling the sheaf of papers she has given him. "Are you going to read us a story?" Brian asks. "No," Vern says, "this is the new fiasco." What he has and is studying appears to be a series of permits. Then this *is* the remodel job he has been talking about. I ask where it is, just to get some kind of fix, and he tells me—"1958 Olive." Jenny and Kevin's house—I know it, it's the house where I signed the lease. I was so upset after they withheld the cleaning deposit that I wrote a pissed-off letter, though I do not remember exactly what I wrote. Still, they remain two people, perhaps the only two people in town whom I dread running into. If I see one of them in the distance, say at Payless or at some community event, I always turn away or pretend I haven't recognized them. They seem to do the same. Now it appears they are the clients for the new job. Something to be added to the humility/humiliation file, although it still hasn't sunk in. The sheer weight of this bizarre twist of fate for some reason helps me believe that I am at the right job and doing what I'm supposed to be doing. What is it about this week—all this spiritual serendipity, with resentments around every corner? Am I supposed to be learning something?

Brian and I end up the day working in the swamp and actually finish the entire wall just short of quitting time. I take over as the angled cuts come into play and there is a nice balance for a while in our working relationship. We talk about religion. I talk about playing football and about having a friend who went on to play at college. Brian asks me where I went to college and I say a college back east. I wonder what he would think if he knew I was a Harvard graduate. I wonder if he has ever heard of Harvard.

I sometimes wonder if everyone has heard of Harvard and at other times I think there are vast pools of the population who have never heard of it. I wonder, out of my graduating class, if I am now working at the most menial job. Would there be some sort of reverse accomplishment in lowliness? And if there were, what would it mean? My father has been dead now for five years, but I think of him now in this context. A Harvard graduate himself, the class of '36, he told me once about a classmate of his who worked as a milkman. To my father, this was an almost un-thinkable abnormality—the only way he could understand it was to believe that the milkman was, in some almost clinical way, crazy. And so what would he make of his son, a $10-an-hour grunt carpenter, *magna cum laude* in European History, covered with mud and sawdust, locked into a long-term resentment with a nineteen-year-old kid from farm country?

Whatever my father might or might not think, Friday at quit-ting time feels no less joyful than it does at the end of any other week, and I drive away in a state bordering on elation. We have been in the rain again all afternoon but my wool sweater has kept me warm. Maybe I'll call Vern up over the weekend and ask for a raise. Yeah, right.

March 13

MONDAY again. At least I thought about calling Vern up. I ran the conversation around in my mind all Sunday afternoon, tentatively planning to make the big call around 4:00 in the afternoon. I pictured it as a mellow moment in the weekend with just the first flickering awarenesses of the week ahead starting to jitter across the screen. I would catch him in a generous mood, dispense with the formalities, make my case for a raise, he would agree, and that would be it—harmony restored to the workplace. But in the end, when 4:00 came I was off buying presents with Emily and Shannon for Kathleen's fiftieth birthday party, and by the time I got home it was already almost 6:00 and I was exhausted. Besides, I figured by 6:00, Vern would not be in a receptive mood to hear anything.

And by this morning, I have arrived at some sort of acceptance about the whole situation and a determination to go sometime soon and look for other work, work with a more compatible crew if that is possible. I can't escape the lingering feeling that I am seen in some ways as the liberal heathen devil on this crew, the enemy Rush Limbaugh is always fulminating about. Thursday or Friday, when we had to move our work station to the far eastern end of the building, I went inside to get the hundred-foot cord and found that it was hooked up to the radio, and that Vern and Bud were tuned in to Rush. I asked if I could take the cord and Vern nodded. Joking, I said to Bud, the real Limbaughmaniac, in the playful voice of a disapproving parent, "You've polluted your mind enough for one day," and he laughed, but I know, too, the serious extremes to which right-wing, God-fearing Rush fans take their views and I wonder sometimes if I am not too flippant about these things. Rush's voice is such a constant sound in the background of this job, but I am rarely close enough to the radio to make out his exact words. Mostly my impression is of a tone, a tone of grating, whining propaganda, a polemicist spreading his gospel of hate through this wide, receptive band of bitter, white males—like Bud, and to some extent, Vern. Bud particularly, short, squat, dark, and brooding—he could be back in Berlin in the thirties, listening to another rabid speechifier summoning the masses to unite against a common enemy.

Today promises more rain. I arrive early and am already in the house when Vern pulls up. We have a quiet moment alone together in the garage. It's the perfect time to ask for a raise and instead I'm tongue-tied and paralyzed with fear. What is it? I project all that authority stuff—the headmaster-cop-father all in one and I can barely say anything except to comment on how nicely the garage floor turned out. It was poured on Friday. Vern's comment is, "At least the little rug rats didn't screw it up too bad," referring to the marauding neighborhood vandal children, the ones who apparently threw the stamped set of plans into the shitter.

Brian arrives and the day begins. We're siding the front of the house now and have our technique down. We work well together. Some hard rain falls early, but by 10:00 the sun is out and we're baking in our raingear. I strip down in stages—hat, then rain jacket, a little bit longer, and I remove my macho-man disco jacket, then the rain pants (my jeans underneath are getting wetter from the sweat), then finally the turtleneck, and for the remainder of the day I am working mostly in just jeans and a tee shirt. It feels good. And we make real progress on the wall.

I leave at lunch-hour to buy one last birthday present for Kathleen—a AAA membership for the stranded motorist. As the end of the day approaches, I know I am going to have to talk to Vern, not about the raise, which I have already given up on, but about taking two days off at the end of next week. I want to go to Spokane to play in an Old-Timer's Hockey Tournament and the bus leaves Thursday night. I need the day, Thursday, to take care of some loose ends. So after work, after all the tools are packed away, after I have already been out to my car and dropped off my belt and my gathered-up raingear, I go back into the house where Vern is on his way upstairs to secure the second-story sliding-glass door. I call to him: "Vern . . ." "Yeah?" He takes a few steps back toward me. I wonder if he is expecting me to ask him for a raise, and the fact that I'm not asking for a raise gives me an unlikely residue of power. I think we both understand on some level that it is unfair that he is not giving me a raise. I say, "I wanted to ask you if it would be okay if I took next Thursday and Friday off." I feel like Bob Cratchit asking Scrooge for Christmas Eve, but Vern's face softens and he wrinkles up his brow and makes himself look like he's puzzling about the pressing de-mands of the job remaining. "Do you think you can get along without me for a couple of days?" I say. "I guess so," he says. "What you got going?" "Oh," I say, feeling vaguely frivolous and absurd. I mean, it would be a different story if I were going after elk or sturgeon or something wild. "I'm going up to Spokane to play in a hockey tournament." He says, "Okay," and we leave it at that.

I drive away elated, already counting the days until my four-day weekend next week. We may be close to finished with the siding by then and the siding is really, except for the mud under-foot, almost fun work. Well, if not exactly "fun," at least pre-dictable and demanding of very little mental energy. I always have a sense of security when I know exactly what my work will be for two or three days into the future. I get home in time to have a cup of coffee before going off to my meditation class.

March 14

My spirits seem to plummet in the middle of the week and today is no exception. The novelty of the first day, Monday, has worn off, and still the weekend is nowhere in sight. We are back to sid-ing, exactly where we left off yesterday; we don't even have to talk to Vern, nor does he talk to us. We hardly see him until break. We are around the back of the house, doing the east wall of the master bedroom and the north wall of the master bath. We make good progress, Brian cutting and me nailing. Brian says he has a headache and feels sick and this slows him down a bit at first. When Vern does appear, he mentions, among other things, that the standard of the industry for six-inch siding is 36 square feet per hour for two men. We start measuring ourselves against the standard and find that we are somewhat ahead. The challenge is now down for Brian—Vern could not have come up with a better, competitive motivator than this—and we com-plete about 200 square feet by lunch. The pace of the standard for one day would only be about 280 square feet, so we feel like we're flying.

Vern leaves in mid-morning to look at the other job. He takes Bud along with him, so I imagine Bud may have some super-visory capacity there. It will be interesting to see how Vern decides to split us up. At lunch, I ask him how the other site looks and he brings out the plans to show me. I feign surprise when I see Jenny and Kevin Blair's name on the cover sheet. "These people were my old landlords," I say, trying to pretend it's all a surprise to me. I have the feeling I've pulled it off, but as Brian would say—"Who cares?" Vern asks me about them and I tell him a few things, not mentioning the squabble we had about the cleaning deposit. He says Kevin seems "Near Eastern or something." I tell him I think that's probably because he's a Jew from New York, but I suspect we're skating into a somewhat slippery racist area there and I don't want to say anything more. Besides, how many Jews would Vern know in his all-white, Christian world? The remodel is small but complicated, and Vern tells me that the truth is he really didn't want to do it anyway; he's too busy as it is. We'll see.

Brian and I have a good afternoon on the garage wall. The work rips along with the exception of one cut on a long board above a window that Brian gets completely backward. He laughs about it on and off for at least an hour, and I think I like him today. He tells me he has been trying to make himself a better worker, and this touches me for some reason, because he is obviously sensitive to my irritations.

All afternoon, I'm working in a long-sleeved flannel shirt with no hat. We're in the sun and I see my reflection over and over again in the window of the garage wall where we are working. I decide I look like an unkempt cross between Bob Villa and Newt Gingrich.

March 15

Wednesday, the Ides of March, and also Kathleen's fiftieth birthday. We are going out to dinner tonight, Kathleen, Emily, Shannon, and me, and then we are going back to Emily's house for the presents. We have bought Kathleen a bird for a present, a blue parakeet in a small cage. Something we think she has always wanted. I'm excited all day, waiting for the big moment.

The day goes so slowly. We hardly make any progress at all. We have only three boards left on the garage wall, but they seem to take forever with tricky angles and multiple cut-outs for windows. When we are done, we move around to the back wall of the garage, where two vent holes and the exterior door-jam require flashing. There are complicated cuts in and around all the obstacles. Things don't seem to be quite level and the "story stick" stops feeling so secure. The story stick is a ten-foot strip of thin wood with pencil lines on it that relate to the spacing of the siding. Anywhere you are on any wall, but starting always either from the very bottom of the wall or the bottom of any particular board, you can always use the story stick to lay out the next sequence. Theoretically, the story stick should keep all the boards level and separated at the exact same intervals all around the house. But there is some fuck-up on this wall and we spend more time messing with it. Our square-footage rate is in the toilet.

Brian starts to drag, to goof off, to joke, to horse around, and this irritates me. And now I think that I have never worked with a more difficult person in my life! Later, we are around on the east wall of the studio, again with a door and a window and a lot of short, complicated, unproductive cuts, and Brian tells me that he believes the most important thing is to concentrate on the work. I agree. I tell him, "That's what I like to do, get into the rhythm of it." This sense of purpose lasts for a while longer before the funny stuff starts again. I like humor mingling with work, but never at the expense of the pace. He likes to stop completely, to sharpen his pencil, to tell me stupid things, and he expects me to respond.

Talk, talk, talk: sometimes I think I'll go crazy from all this chatter, and I look longingly across the job site to where Bud is quietly and peacefully caulking away, his caulk gun, his rag, his little one-inch putty knife. Alone. He's been listening to Rush again, quietly and peacefully filling his head with all that venom.

He remains an enigma. Vern is around all day with one sub-contractor or another—the security people, the phone people, the concrete-saw guy arrives for an hour to cut the channels out for some of the plumbing and gas hookups. It's a five-ring circus, and Vern has few opportunities to check on us. Still, he can always pop up at an unexpected moment, come around the corner with his tired face and his "How's she goin', boys?" and then head off again to some other spot, some other problem.

The day clears up in the afternoon and we have sun, the grass on the golf course is a brilliant green, the golfers come by in a steady procession, and we pound up those siding boards, one slow board after another. After all these days, we are still only halfway through the siding, if that, with a lot of the high, inaccessible places left. I leave on the dot of 4:30 to go get ready for the party.

March 16

Another day of siding. We move from the back wall of the garage to the big side wall, the highest wall in the house. We will soon be up at the farthest extension of the biggest extension ladder, a place where I feel less than comfortable. At the top of one of these big ladders, you are standing on only one rung and usually you have your body pressed as much as possible into the deceptive protection of the wall. From this position, you then reach up cautiously with a nail in one hand and your hammer in the other to secure a board that might be a foot or two over your head. Particularly on the fake pressboard siding, just to get the nail started is sometimes a major operation. Soft and synthetic as it is on the inside, the siding has a tough, almost plastic, exterior skin. But today we are still down on the lower rungs of the big ladders, cutting thirty-degree angles into the left end of each run of boards. This is one of the walls that directly faces the line of fire of the golfers, so there are no windows at all. A huge expanse of gray siding, probably about twenty-four feet high at one end, tapering down to ten feet high at the other, and on a run of about twenty-five feet at the longest point; by my calculation something like 325 to 330 square feet. We still have a ways to go.

Brian is planning to leave early, so he works through lunch. Bud has gone off and I am left again to have lunch alone with Vern. We are sitting outside in the sun on what remains of the pile of siding. We talk about Carol Anne, his grandchild with the tumor. "We had a party ourselves last night," he says. "A fifteen-pound party." This in response to my mentioning Kathleen's fiftieth birthday. The fifteen pounds confuses me. I think, is he referring to Weight Watchers? Had Connie, his wife, lost fifteen pounds? No, it was to celebrate the fact that Carol Anne had finally made it to fifteen pounds. This has been one of her main symptoms, this inability to gain weight. It's a hopeful sign, but for the rest of lunch we talk about the fearful prospects of her tumor, its mysterious nature, its uncertain diagnosis, the doctor's fears that it may suddenly "go hot" and flare. I can hardly imagine what the daily anxiety must be like for Bud and Debbie.

At 2:00, Brian leaves to go with his wife for a prenatal checkup. This will be Brian's first baby, and apparently it has already been identified as a girl through sonograms.

I am left to work alone, the first time I have worked alone, it seems, in weeks, maybe months. I move over to the north wall of the studio, a one-story wall with four large windows facing the golf course. I have my cutting table set up in the sun, but am working around the corner in the shade. The weather has become almost balmy, sunny with a few patchy clouds, temperature in the high sixties, golfers drifting by in regular foursomes. The work is steady, easy and absorbing. I do all the cutting right here, don't bother to go around the house to the radial arm saw, because I want nothing to intrude on my solitude. The hours are profoundly enjoyable and remind me how much I love to do carpentry. On a day like this, in quiet, in peace, with no rain, with the boards lining up easily and falling into place, the nails plunking home, one after another, no ladders, no mud, a boss who comes by maybe once all afternoon, and then just to say, "Well, how's she goin?" I am in a state of perfect flow, of harmony, of almost mindless happiness. The work, itself, becomes a prayer.

The two hours alone force me to wonder why it is that the universe has arranged for me to work all the time with Brian. It must be some process of humbling, some test of love and forgiveness that I am expected to work through. A spiritual exercise in compassion or mercy or just loss of ego.

As if to add to the perfection of the afternoon, Vern comes around ten minutes early to say we're knocking off so we can go watch the big NCAA tournament game. Tip-off is at 5:00. Back at home, I sit and watch most of the game but care nothing about who wins or loses; it is enough to sit and be quiet and be at home and not have to do anything for a few hours.

March 17

St. Paddy's Day, although no one mentions it. I don't think Marshall is an Irish name anyway. Brian and I finish up the wall that I started yesterday afternoon and then move around to the other big wall, the wall of the studio, almost as big as the big wall of the garage, maybe 300 square feet or more. Again, every left-hand cut has to be a thirty-degree angle to fit up snugly against the soffit plywood under the eaves.

As a joke, Brian bought me a hammer at the Dollar Store in the mall last weekend. It's a little, tiny hammer with a short, wooden handle and a curved claw, almost a toy, or the kind of hammer a bachelorette might own to bang a small nail into the wall of her apartment so she can hang up a picture of seagulls perched on a piece of driftwood or Telluride in January or Trumpa Rimpoche. Certainly not a hammer for a real, macho carpenter like me with my big, booming "California Professional," the 23-ouncer I bought with Vern's Fred Meyer Christmas gift certificate.

For most of the siding job, I have actually retired the California Professional and have been using "Johnny One Claw," a small, 14-ounce hammer with a red wooden handle, a rubber grip, and a broken claw, the claw that broke over at the Baxter job when I used the hammer to yank out a stubborn sixteenpenny nail. With one claw, it is even lighter than fourteen ounces, and still it works perfectly for nailing up this siding, and gives almost no strain to my wrist. I have even stopped wearing my elastic wrist support.

But now, up under the eaves, we discover a bit of a problem. The fascia board is only nine inches from the wall of the house. It is also nine and a half inches deep, and so gives you almost no room to swing a hammer. But the tips of our angle cuts have to be nailed off or they'll be flopping out in the breeze like so many curly shingles—hardly the look we're after. I go to my car and get the little Dollar Store joke hammer and it works perfectly. Johnny's one claw is a straight claw and not only gives you no room to swing but also seriously gouges the inside of the fascia board, not a high area of visibility but still . . . This little hammer with its curved claw is just the ticket. As we go up the side of the wall, I am amazed with every board at how perfectly the little guy does his job. I'm calling him "Johnny Dollar" and sometimes "Lillehammer," after the Olympic site in Norway.

We work away at the wall with very little tension. The siding has become such a process that we seem to be working more in harmony than ever before. And the house is moving ahead so rapidly. I wonder how much longer we'll be here at Chin's. Today the security people have been installing the alarm system. The installers are electronic types—clean like electricians but more marginal, somehow. The boss is a tough-looking character with a Southern California vibe. Big ears, slick hair, he wears formal shirts with rolled-up sleeves and jeans, has a beeper on his belt, but these days who doesn't? Me, for one. He looks like he might have been a cop. His installer has a mellower aspect, wears a new-style/old-style baseball cap that has no high crown like the regular baseball hats. He may be a hiker, a cross-country skier, a tree hugger, an outdoorsnik of some kind, maybe a kindred soul for me on this job; we have a few short conversations in passing, mostly about the local minor-league baseball team. The telephone people are here too, running wires like crazy. One of them arrives in a small car with a sport rack on the roof, room for bicycles, snowboards, whatever. He also has one of these new-look hats and he practically runs around the site, stringing wires.

Brian comes back from a trip to get nails and is tickled with delight because he had an exchange with the telephone guy in Bud's presence about Rush. Rush was on the radio as usual and Brian asked the guy what he thought of him and the guy said he thought he was a "fat fuck and full of shit," which makes Brian chuckle for about two hours. As so often before, a light-hearted mood carries us through Friday afternoon.

I have been taking my watch off or leaving it in the car to try to keep from looking at it. 10:56. 2:05. Seeing these dead hours in the middle of the day demoralizes me. Now, this afternoon, I put my watch on, the better to savor the slow pace of the last hour and a half of the week. The sun has disappeared. The clouds roll in. A few sprinkles fall and the air is cool and fragrant with the budding flowers of spring and the moist, freshly cut grass of the golf course. I am happier and happier as the final minutes of the work week tick by.

On my drive home I think, if you could only bottle that Friday after-work feeling and sell it to people, you could make so much money you could stop work and then you would never have that Friday after-work feeling again. Unless you indulged in your own product. And probably, after a while, you'd get addicted to it, it would lose its kick, it would turn out to have negative side-effects and all would be lost and in ruins. You would lose your fortune and have to go back to work and then some Friday you would be driving home and you would have that Friday after-work feeling all over again.

March 20

AND Monday again. My spirit feels broken by this job. Not just the people but the house itself seems wearisome and detestable. Are we getting anywhere? I turn down Cypress Loop, take a left on Cypress Way, and there it is, alongside the golf course. I'm five minutes late, the first worker here; Vern is already in place with his white truck. When I come walking up the driveway, he emerges from the garage and I ask him for the key to the back of the truck so I can start unloading the tools. Every act seems hopeless and futile. I start by throwing the power cord toward the power pole, except I hook it badly into the wood pile and then realize I have thrown the wrong end anyway. Vern is watching. Of course! I wonder, once again, about what he thinks of me, that I am some kind of clown, some fuck-up barely worth the $10 an hour he pays me. That if he could think of some gracious way to lower my wages, he would. It's the day before the first day of spring and my heart feels hollow and dark with despair. With a desultory and hopeless rhythm I try repeating my mantra to myself. I set up the ladders on the back wall of the garage and start trying to hang the siding myself.

Brian shows up half an hour later, full of excuses. I can't stand him! The sight of him fills me with a loathing so deep I have to start praying with redoubled energy. As we hang the first few boards of the day, he gives me a running replay of a movie he saw over the weekend, some made-for-TV thing about a woman who shot her husband, cut him up, and kept his pieces in the freezer in the basement. I get all the scenes back one by one over the course of an hour and a half and I never have to say a word. Later Brian gets going on that song, "In the jungle, the mighty jungle, the lion sleeps tonight . . . oooohoooooowhooohhoohohh . . ." and sings it over and over and over and over again.

The day is cold and blustery. After lunch, when we have moved around to the big garage wall, the wind comes up and then the rain and then the hail. Suddenly it's so cold, my hands are numb even inside my gloves. I have to get into my full raingear. It's as if winter never left.

Up on the ladder, I feel shaky and somehow out of my body and I wonder for one paranoid moment if someone might have slipped LSD into the Dari-Mart coffee container. I feel almost that strange and almost that dizzy. I think of just taking off my belt and walking away from the whole thing.

This past weekend, Kathleen and I spent a night in a motel out on the coast. We stayed in bed the whole time, eating, reading magazines, watching TV, making love, and sleeping. As we were leaving on Sunday, I said that it was as if we had spent twenty-four hours in a hospital. We never even went out to the beach, just lay in our bed and stared at the angry ocean through our picture window.

Work has become like a scourge, a plague, level three of the *Purgatorio*. And the siding seems to go on forever, the siding of Sisyphus—is that what this has become? At the end of the day, we are back on the front wall of the garage and moving slowly up. Vern is around setting corners. Bud is doing a slow, meticulous job on the small siding pieces that have to be set into the multifaceted front walls of the kitchen and front entryway. Brian

continues rolling along in his usual obnoxious way, bossy, demanding, controlling, childish, and every once in a while just
charming enough to get me at least to talk to him. I can't wait for
the day to be over. The cold is supposed to get worse.

Someday this house will be done, and we and all our voices
and our inane chatter and our grim faces will be the ghosts of
the workmen who built it, and whoever lives in it will know us
only by the strange, unsettled sensations of movement and
thought they pick up sometimes when they squeeze into the attic or lean against the far corner of the garage wall or look at the
roof or rub their hands over the outside walls of the master bedroom. We will leave the faint outlines of our troubled selves behind us. We will have moved on somewhere else with our
yearning laughter, our muddy boots, our hammers with their
odd nicknames, our ladders, our wet nail-belts and tired feet,
our fear of fuck-ups and death. And the house will be like our
mountain and it will remain.

March 21

First day of spring? The weather is moody and unpredictable.
We have cold winds at first and bursts of bright sun, then showers, then hail, then dark clouds that boil in from the west and
drench us.

The backhoe is here digging trenches for the electric, sewer,
and water lines. We race to stay ahead of it, to finish the walls we
have to finish, and avoid the nightmare of trying to work off ladders set between a three-foot trench in the ground and a three-
foot pile of dirt. The backhoe is fast, but we are faster, as we scale
the top of the west garage wall and then tackle the north wall that
rises up and crosses over the small, flat roof over the stairway.

Brian becomes totally bossy and unpleasant, and I act out his
orders with some speed just because to keep moving is to stay

warm. At one point, he gets a dark smudge on his upper lip, either from his pencil or some dirt from the boards, and I tell him
about it. "There's a smudge on your lip," I say, and as he starts to
rub it off I add, "a small line, like a little mustache, actually like
an Adolph Hitler mustache." I am suddenly struck by how perfect it is that at this moment, as he takes on his full monster/dic-
tator persona, he has sprouted this little Hitler mustache. He
wipes it off. I can't tell if he gets the reference. Herbert, the old
roofer, the ex-paratrooper, the guy who was in that first wave of
infantry to land in France, overhears us and kicks in, "Oh, he
don't probably even know who Hitler was anyhow," and we have
a link there. "Yeah," I say, "he thinks he might have been one of
the presidents." And Brian, who has no sense of humor about
himself if there is any implied taint of stupidity, says he knows
perfectly well who Hitler was and the conversation is over.

He takes all the hard, high jobs at the top of the wall because
I can't reach it even when the ladder is extended to its last extension rung, but I am grateful too that he is so much bigger that
I don't have to work twenty feet up with my body pressed against
the wall and the hammer up and out at arm's length. If I did have
to do this job alone, I would have to build a scaffold or platform
of some kind to set my ladder on. And I would do it. One thing
this job has taught me is that I am not Brian. I am a fifty-one-
year-old man who is only five foot seven, and I have some limits
of fear and caution beyond which I won't go. Period. If Brian has
to say, or even think, "What's the matter—leave your guts at
home?" so be it.

By the end of the day, we have completed nine of the fourteen
or fifteen separate outside walls of this house. When I stand
back, across at the turn-around where my car is parked, and look
at the house, I can see it in something of its finished form. As the
gray siding replaces the white and tattered Tyvek, and the last
panels of roofing drop into place, the house starts to take on a

trim and solid quality. Yet, in spite of that finished look, I am struck all the more by the cheesiness of the materials. The foundation is concrete and the studs and beams, the real bones of the house, are real wood. Vern has talked from time to time about the quality of the wood, which is all small growth, second- or third-growth harvest with very little of the old-growth fir he built with when he was younger. More knots, more bark scars and ragged edges, less clear grain. But that isn't it either; it's the thought of all that pressboard plywood and the pressboard siding with the imprinted grain, a repetitive grain pattern that is supposed to make it look like real wood. To me, it looks about as real as kitchen floor linoleum printed with renderings of brick. And the roof, which is a stamped metal product that is made to look like ornate Mediterranean tile. And the cheap vinyl windows and sliding doors, the metal exterior doors with the fake wood grain, the plastic pipe, and on and on. When the final details are added, when the marble-stamped Formica and the fake-grain finished plywood and the polyester carpets are added, the overall effect will be achieved—the illusion of grand style and meticulous workmanship, bourgeois splendor acquired on the cheap. If there was an antithesis to the Shaker aesthetic this would be it. Form follows ego. As I stand by my car at the end of the day and look at this thing we have built here by the golf course, I think not of great old houses and fine old-world craftsmanship, but of fake, ticky-tacky, pseudo-fancy products, and how everything seems ultimately bound up with illusion. We make it for less, but we make it look like for more. The house is a stage set and the play is all about power and luxury for everyone, power and luxury at an affordable price. Five-bedroom house—fully loaded.

March 22

This then is my Friday. I am going to take the last two days of the week off and go to Spokane to play in the "Old-Timers Hockey Tournament"—a chance to spend four days with this group of old guys I play hockey with all week. I'm a little uptight about the drinking, but figure I've been sober long enough (thirteen years) to withstand the temptation. Mostly I'm anxious about feeling like a nerd. But what else is new? I feel like a nerd most of the time on the construction site anyway.

It's another day of siding; I think this is the twelfth or thirteenth day in a row that we have been on the siding. You can only go so fast, I keep telling myself. Brian and me—the perpetual team. Up on our ladders, we make our marks an inch and three-eighths down from the top of the board, snap a line, measure, cut, nail, re-mark, and repeat. When we are separated by any distance at all and up on ladders, we have to do the chalk-line toss. We use Brian's chalk box because he's Brian and it's his and that's the way it is. He has blue chalk. He pulls out fifteen feet of line and throws the line with its little metal clip at the end. This is sometimes hard to see, but it is light and harmless. The worst thing that can happen is that I will miss it and the string will fall into a puddle. Brian is almost always the thrower and I am the catcher. Sometimes I make the catch, sometimes I don't, but today I'm hot, I can't miss, I snag anything he throws. He's throwing to my left hand, which is my glove hand, but that doesn't explain it. I'm in that perfect catching space.

We move ahead for a while on the front of the garage and then get caught up in some complex angles around the patio door and the roof above the small guest bedroom. Brian takes charge, does all the figuring and cutting, screws up everything, and then wants me to fix it. I hate working with this guy, and a part of me seems to see suddenly that it can't go on much longer, that I am indeed going to have to give notice and walk away from this job.

[163]

Today, I think that my regular carpentry job has been compounded by the task of baby-sitting a hyperactive child. "Have you ever been formally diagnosed?" I ask him later in the afternoon, when he's singing his inane Christian children's songs about how to get along in life.

One song in particular is called something like "Nathaniel the Grublet," but I tune out the words. A few nights ago, I had a dream that he was singing this song to me, and I was screaming, "I hate that song, I hate that song!" But I am beginning to realize that as crazy as Brian is in his own little innocent and insidious way, I am becoming that crazy myself. In fact, I realize, the longer this goes on, the more obvious it is that the problem lies with me, and not with Brian at all. As we say in AA, anytime you're pointing the finger at someone else, three fingers are always pointing back at you. I am constantly sucked down into the pit of this neurotic duality, and again I am convinced, as the afternoon drags on toward quitting time, that I will have to quit for good. Even with Brian gone, I think I would have to move on to another place, a place with more compatible people. Funny— the problem with Brian is that he laughs and chatters all the time like a squashy-headed town bumpkin, while Vern and Bud mope around with their eyes cast down and their thoughts to themselves. This has to be some sort of test, but what am I being tested for, and how much longer can it go on? Hard as it all is, I am not, I realize, a quitter. With a sense of resigned finality, I have to face the fact that I am going to ride this one through to the bitter end. And maybe, just maybe, the bitter end is closer than I think.

Like a passenger on a ship just off the edge of the horizon, I have this instinct that I am approaching land, but I still don't see it. Tomorrow, in my absence, they are going over to Jenny and Kevin Blair's house to start the concrete demo work, so I'll be over there soon enough. The time at the Chin house is coming to an end. The inspector has signed off the framing and the insulators are coming in at the end of the week. I don't know what more we'll do. The rest of the siding, I presume, but beyond that, maybe nothing. "That's all finish work," as Vern said, and I don't think he sees me and Brian as "finishers." For now, it's the end of the week for me and I'm off to Spokane to play in a hockey tournament.

March 27

MONDAY morning and I'm back at work after the four-day weekend: twenty hours on a bus, three hockey games, trips to rowdy sports bars in the role of the designated driver— a good, long time to get away from the immediate drudgery of work. But I face this morning with dread, and entertain notions of giving Vern notice. I run the words around in my head: "Vern, I'm going to be leaving in two weeks." "Oh yeah, what you got goin'?" Should I lie about it? I could tell him I have a big, new job that pays $21.00 an hour or . . . I don't know. I'm so paralyzed by the father-son transference, so threatened by his authority, that even the thought of telling him I'm moving on is almost nauseating. Calling up a girl to ask her for a date at age thirteen is about the only thing I can compare it to.

The day is bright and beautiful. I get to the site fifteen minutes late and Vern and Brian are already there rolling out the cords and tools. We go right back to work on the wall we left last Wednesday. Apparently they worked at the other job, the remodel for the Blairs, on Thursday and Friday, breaking out concrete and forming up for the footing for the new addition. Nothing changes as we get back up on our ladders, measure, snap lines, and slowly move ahead up the wall. This is our sixteenth day on siding. Vern had ordered about 5,000 square feet, which should take us roughly twenty days to lay up, so I guess we're not too far behind his projections.

My attitude is better. Brian's chatter seems familiar but happy, and I enjoy being back in place, cutting and fitting boards, mindless in one sense, mindful in another. The golf course is a warm green and flooded with sun-happy golfers who come by in regular shifts. All day we are on the golf-course side of the house, and Brian runs off periodically to comb the field next door for lost balls. A warm-up for Easter.

The only glitch in the day is the stupid roof, this crappy metal material pressed into the shape of Mediterranean tile; you have to walk on it in just the right way or it will buckle along the ridges of the fake tiles. We are siding the gables on the north side of the house, working off the roof and not the ladders. Before we have made it halfway up the first gable, the small eastern wall of the guest bedroom on the second floor, we have punched in three tiles. They are not badly noticeable from the ground and not broken (they will not leak), but still, to a practiced eye there is a definite skip in the regular pattern of the course going up the grade. It will not do to have the roof all squashed down under every gable where we work; we have two more much bigger gables to do after this one. I hate this material. As if it weren't hard enough to be up on a roof at all.

Still, the day is peaceful and easy, good warm weather, and Vern is gone until mid-afternoon, away at the other site waiting for the inspector, so it's just me and Brian. A bedraggled insulation installer is around for a while and then leaves, claiming he is suffering from the after-effects of a bad bar fight. Bud is away in Portland with Carol Anne, who is having her third chemotherapy treatment. Tomorrow we are scheduled to pour the foundation at the Blairs' in the afternoon.

As if to tweak my resolve, a woman came up to me this evening after our meditation class to ask me if I would give her a bid to build a huge geodesic dome, a duplex she said. Greg was there with me, a new friend and also a builder and cabinetmaker and sometime-architect and so, against my better judgment (or really, against my fear), I took a set of plans to look at. Is this the Universe's way of showing me the road I have to take to get away from my gig in Purgatory? Is this the road to peace or the road to major stress? The plans look appealing in some ways. We'll have to see.

March 28

Sunny, bright, clear, the temperature moving close to seventy, the most beautiful day of the year, the flowers out everywhere, the trees blooming, that soft, fragrant essence of freshly cut grass lingering and swirling like some constant, steady vapor cloud over the city. Moving as slowly as one of the few lazy clouds that meander inland from the coast, we start out, back on siding again.

Brian seems goofy and uninvolved, but counters any suggestion I make about what to do or how to do it. The most annoying person in the history of work! But today I don't give a shit. He tells me about his new class this semester, a class in management techniques, and I talk for a while about that book *The One Minute Manager*, which I never read but which I heard about from somebody. The essence of it, as I remember, had to do with positive reinforcement. Vern is not around for part of the morning, so Brian and I talk freely for a time about his management style. I tell him about the one time Vern praised me for my skylight job. And, for a while, we are bonded again by our common position as Vern's employees.

To counteract the disaster of the crushed roof tiles from yesterday, we build a temporary, hooked platform to fit over the ridge of the roof we're working on, and this takes up more time. We have to Tyvek the rest of the wall, so we have to find the studs on the gable truss, which are not laid out like anything else; some are at odd angles, some are at fourteen inches, some at sixteen on center. The process takes forever. When he returns, Vern comes out with an electronic stud finder, and even then we pound in nails all over the place and find nothing.

After lunch, we break away to go and pour the footings for the remodel at the Blairs'. I'm not there ten minutes when Jenny shows up around the corner with a bright smile and says, in full earshot of Vern and Brian, "Where's Joey?" in a cute, innocent, sing-song voice. "I hope he's not still mad at me." It's a little obnoxious the way she says this, but I'm glad she brings the whole thing right out into the open air and we chat about things. "That's all water long since under the bridge," I say, referring to the cleaning-deposit dispute, and this seems to ease the residue of tension. Even so, I monologue for a while with myself, go back over the same old ground, and still feel ripped off of my two hundred bucks. Well, it truly is water under the bridge, and my little resentment doesn't last for long.

The concrete truck finally shows up, and we hump two yards of concrete in wheelbarrows from the front of the house. This is a small-scale remodel, and again I wonder how much longer the work will last. Once this is done, there's no telling where we'll go next—if anywhere. Maybe I won't have to quit at all. Maybe I'll get laid off and go on unemployment! I can only dream. We finish with the concrete at exactly 4:30. In the open air, on a clear, warm day, doing heavy labor with Brian, having a lot of quiet time with each wheelbarrow load, saying my mantra, daydreaming, dry and warm and comfortable, I think as I drive away at the end of the day—hey, this job's not half bad.

March 29

Another beautiful spring day. We show up at the Chin house as usual at 8:00, and find the sheetrockers already hard at it, two grim, determined old-timers, like a couple of bikers just back from a long run; they are doing the ceiling in the front room. The insulation is already in place, the painters have been here and primed the fascia boards, the last of the outside work is being wrapped up—in fact we are among the last ones to work on the outside. The siding, the endless siding.

Bud is not with us, so it is just me and Brian and Vern again, and in the new three-way dynamic Brian and I have become like the odd outsiders to Vern's one silent, stern, cheerless boss. He says very little, even at lunch, but sometimes I think it's because he can't squeeze a single word into Brian's continuous monologues. Today we get into a discussion of bugs and snakes, and Vern finally opens his mouth and mentions that the thing he hated most in Vietnam was the snakes. It's a perfect opening to ask him more about what it was like over there, but I hesitate. Here it is again—that rift that runs right down the center of our generation. He doesn't want to know what I did about the war in Vietnam either. The hippie and the soldier, the pacifist and the baby killer, the traitor and the patriot—in our minds it's still Berkeley, 1969. Brian says something about how he's glad he doesn't have to listen to Rush Limbaugh all day, and Vern says, "Well, he's the only one around who speaks the truth . . ." Says it loud, like he wants to prod me into a response, but I bite my tongue. I certainly don't want to get into one of those arguments! Still, the rift is there between us, and Vern seems more morose than ever—perhaps not having his son around as an ally.

Brian and I keep hacking away at the siding; we are in a complicated area where a fifteen-degree porch-roof overhang leads back up into the thirty-degree roof, and the work drags interminably. Brian is down cutting at first and sharing the Skilsaw with Vern, who keeps changing the depth of the blade. A few other things happen, words are passed, I'm not sure what exactly, something about Brian having to go get his own sawhorses and set up his own work station; whatever it is, it has Brian totally pissed off. He says, among other things, that he now understands why Bud is the way he is because he's had to put up with so much shit from Vern for all these years. He starts talking in a fairly loud voice about all of this, even while Vern is right up above us, somewhere on the roof, sheeting the chimney box.

The cabinet guy comes around for a while to look over the cabinet situation. The electricians are here. Some other people drop by about something. Vern seems to jump constantly from place to place. Brian and I roll along. Brian begins to surface out of his bad mood.

Up on the ladder, standing in the sun, waiting for Brian to repair one of his many fuck-ups, I have a minor epiphany; I'm saying my mantra and I have a moment that verges on joy, a free-floating sense of impending ecstasy, a glimpse of the truth—and it has nothing more to do with Rush Limbaugh than it has to do with the old hunks of discarded insulation that I am staring down at from fifteen feet up. The insight has something to do with now, with time, with a fixed point, a perspective of death. As I drove to work today and passed the decapitated body of a road-killed possum, I thought—that's okay, the soul of that possum has moved on, maybe to a higher, a newer, a better, a non-possum existence. And I am thinking something similar this morning from the top of the ladder.

Greg and I have been talking about forming a partnership and putting in a bid on the dome job, but the prospect, while attractive in many ways, brings up all kinds of anxiety for me. Maybe I'm just not the managerial type. Maybe I need a boss. I need to be told what to do, I need someone else to take on all that responsibility.

March 30

And yet *another* perfect spring day. We start at the Blairs' house, forming up the stem wall onto the top of the footing we poured Tuesday. The total stem wall is only about sixty feet, a few turns, a few right angles, eighteen inches high and six inches thick; the whole project takes us four hours and we are finished in time to go back to Chin's for lunch. This work is becoming almost routine in its moves. We all seem to know exactly what to do. Vern gives us no more than minimal directions on the stem wall; the process involves plywood, rebar, clips at the top and bottom, a two-by-four runner along one top edge, and a few tweaks to straighten out the wah-wahs in the old formboards, which are used for job after job after job.

Jenny appears in the early morning to show off her sixteen-month-old son, Jackson. He's almost exactly the same age as Carol Anne, and must weigh twenty-five to thirty pounds, with a full head of red hair and a shy but penetrating gaze. I guess Vern picked up on Jenny's remark about "hoping I wasn't still mad at her," because he asks me at one point in midmorning if there had been some trouble when we moved out of her apartment. I say, "Oh, just a little dispute about a cleaning deposit—no big deal," and I don't elaborate. It seems pointless to dredge up that old resentment. In fact, I am happy to have the chance to make friends with Jenny again, as I always liked her and hated skulking around town with this vague fear that I would bump into her sometime at the deli or at a basketball game or something.

Back at the Chin house in the afternoon and back in the siding saddle again. The process is becoming so routine and boring, it is hard to keep any kind of an edge. Brian is dragging his ass through most of the afternoon, far more interested in the golfers that come by than in the endless rows of boards that we continue to slap up on the sides of the house. Vern goes up on the roof to mess with a demonically tricky little gable segment created by the guest bedroom wall poking up slightly higher than the roof of the studio. He is up there for most of the afternoon siding an area that could not be much more than fifteen to twenty square feet. At one point, he calls down from up on the roof, "I almost became a statistic!" "What happened?" we ask. "A golf ball landed about two feet from my head!" he says. Sure enough, a few minutes later along comes this old duffer looking for his ball. When confronted with the fact that he almost beaned Vern, he seems blithely guilt free. "Oh," he says, "I knew that was gonna happen sooner or later. I was just hoping it wasn't me that did it." To a golfer, anyone who builds a house alongside a golf course is crazier than a bedbug anyway, and deserves whatever scars, nicks, and dents they pick up. As carpenters building the house, we are regarded as a species just a few notches up from lunatic.

The house is coming along, and we only have a few more walls to go before the siding is complete. Presumably we will then move over to Jenny's, but it's hard to know how long we will be occupied over there. Maybe Vern is planning to lay us off. I can only hope.

March 31

A cloudy, cool day with small episodes of rain. Bud is back from Portland, and we are still on siding, although the light at the end of the siding tunnel is now clearly visible. We finish up the big, complicated west wall of the main livingroom, and move at the very end of the morning over to the east wall, which is, I think, the last unsided wall on the house.

Vern is up on the roof above the garage finishing the two small dormers and we hear him cursing and fuming from time to time; one time in particular, he loses something off the roof as he is climbing up, either a tool or a piece of corner stock, and we hear him shouting at it in that sputtering, irate Donald Duck voice. Brian and I talk about this. And I say to Brian, "Vern may not have given me much positive reinforcement over the past six months, but I will say, too, that he has never cursed me out and I'm grateful for that. He's cold and uptight and rarely expresses anything, and he seems to hate his work, but he doesn't take it out on the employees." And Brian says, "That's right. I used to work for a guy and he'd yell at you, 'What the hell do you have to cut that board three times for?' and shit like that."

If I were working for someone and doing an honest job, the best job I could do, and they started yelling at me like that, started yelling at me the way Vern yells at boards and tools, as if they had a willful and uncooperative mind of their own, then I would simply put my hammer back into its hammer sling and walk off the job. There's no excuse for talking to employees that way and I thank God that I don't work for someone like that; I thank God I'm not one of Vern's tools, or that piece of wood he dropped, for that matter.

After lunch, we head over to the Blair job to pour the stem walls. The concrete truck is already there when we arrive, and so is Bud. The job goes well, quickly and efficiently and in an atmosphere of mutual appreciation, with Brian and Bud wheeling the big, heavy wheelbarrow loads in from the street, me directing the wet concrete into the forms, tamping it and rough-screeding it, and Vern coming along behind doing the finish troweling and placing the J bolts. It's all so routine somehow. I enjoy the smooth, easy pace of the work.

We finish up about half an hour early, and at the very end help Brian load a huge, discarded piece of garbage cabinetry onto the back of his truck, a hunk of junk Jenny decided to throw away and Brian decided to take for his new house. A light-hearted Friday afternoon mood prevails, and there are lots of jokes about the way Brian has tied down the shelf unit to the back of his truck. Bud imitates a traffic reporter: "We have a load of what looks like cabinet components in the number two lane of Highway 149 and that has you backed up all the way to Wal-Mart on the northbound side . . ." We're off early and it's the weekend. Beverly Chin's is closer and closer to completion, at least the framing and siding parts of it that have been ours. I don't know where we will go after the Blair job. Ever the cowboy sphinx, Vern offers no clues to the future. I secretly hope I get laid off. I don't know how much more of this I can take.

April 3

ANOTHER Monday at the Chin house. Vern is already caught up in the demands of the Jenny Blair remodel, and leaves Brian and me alone after the first hour to finish up our siding. Except for the small dormer on the garage roof and the two glass-block walls that flank the front door, all the siding is done after we finish this wall. We are on the east wall of the main-bedroom side of the house, a large gabled wall that goes up and around the eastern-most point of the house, a small, roofed extension that houses the master bath and guest bath. For most of today, we are working from the roof of this structure, cutting and fitting the angled siding pieces into place. Except for the two sheetrockers who hack away at a relentless pace indoors, we are the only two on the job.

In the morning, as Vern is preparing to leave, he comes out to where we are working and asks me to follow him. I walk behind him around the side of the house, and he takes me indoors with a look of some serious intent, but as it turns out, he only wants to tell me that there are several things that need to be changed if the electrician should stop by. A plug in the bathroom, for example, is right in the middle of the mirror. And a motion detector spotlight needs to be mounted on the west wall of the garage.

He leaves, and I go back to Brian, who thinks I have either been fired or given a raise. His assumption is not off-the-wall—it did seem as though Vern wanted to take me out to the woodshed. But he so often gives off that impression—the dour, cheerless leader and taskmaster, forever dissatisfied, forever stern and critical, convinced over and over again by what he sees that it is indeed hard to find decent carpenters anymore.

After Vern leaves us alone, as if to act out his apparent prejudice, Brian and I goof off for a while, go slow, mess with the templates and angles for the boards we are cutting, which are supposed to define a perfect, straight line up the roof. The line is less than perfect, but it's hard to know how much less than perfect is acceptable. Whose eye will be looking at this work? How do you justify the time it would take to make that line come out perfectly? And perfect to what—a sixteenth, a sixty-fourth, a red cunt hair? Brian has a way of saying, "It's close enough," and I have a way of going along with him, thinking my worst resentful thoughts, which are variations on the theme that since Vern is paying me only 50¢ an hour more than he is paying Brian, then he will only get from me 50¢ an hour more work. It's my passive-aggressive instinct.

The day is beautiful and warm and by early afternoon we are working in the shade of five big cherry trees that are in full bloom. Every trip up the ladder, I brush against the thin, outstretched boughs laden with blossoms. By 1:30, we have gone into high gear, and we race up the other side of the gable, meet our first side within a quarter inch of level, and then climb the last few boards toward the peak.

Vern comes by to check on our progress at the end of the day. We fall only three boards short of completing the wall. I tell him I will be late tomorrow because I have to take Kathleen to the airport. She is going to L.A. to continue the celebration of her fiftieth birthday with her sister and some other female friends.

Vern tells Brian to come in early. It will be time to start picking up the scraps.

April 4

Tuesday. I come in two and a half hours late after dropping Kathleen at the airport for her flight to L.A. The weather has changed again and the rain is falling when I pull up and park. No sign of Vern's truck. I find Brian out back in full raingear and covered with mud. Our job for the morning is to pick up all the debris around the house. The mud is back, as thick and sticky as ever, and I put on my own raingear for the job. Most of the debris is what we put there earlier, when we were sheeting first and then siding in the rain—hundreds of plywood and cardboard scraps we laid out on the mud to hold up our ladders. Now all this stuff has sunk down in the alluvial remains of some earlier geological age. With gloves on, boots, rain pants, rain jacket, and black sponge, I start hauling the larger, water-logged strips of wood around from the back of the house to the scrap-wood pile next to the trailer. I am soon as covered with mud as Brian. Brian concentrates first on the old skylight shipping boxes, which have turned to great, soggy brown sheets of mud-saturated cardboard. We haul wheelbarrow load after wheelbarrow load out to the small wooden trailer that Vern will then pull to the dump. And now I really begin to feel the end of my involvement here. We may come back to do some small pieces of finish work, but somehow I doubt it. And I take the time now to look at these boards we are picking up, to remember the things I see, the old blue string from Brian's chalk box when it broke, empty tubes of caulking, the odd scrap of two-by-six or two-by-four left over from the framing process—though most of this has been hauled away in an earlier load—the ribbons of Tyvek, the pieces of black paper from when we covered the roof, the water-swollen hunks of particleboard

from the wall and roof sheeting, the many-angled butt ends of
siding from our last long chore; all of it together gets loaded into
the trailer and by lunch-time we are almost done. Our last arm-
loads come from the front, from the piles of discards under the
radial arm saw, most of them no bigger then six or eight inches.
Soon, perhaps this week, the yard will be graded smooth, and the
raw, uneven terrain of the construction site will be buried under
fresh soil and sod. There will be azaleas and rhododendrons and
flowers of all kinds. No more glimpses of the footings; they will be
a foot or more underground.

At lunch, we sit one last time in the dining room in the dust
that hovers in the air from the sheetrockers' many cuts. The
sheetrockers join us. We sit on piles of sheetrock and talk about
pregnancy and welfare and I tell them that the best year I ever
spent was the year I collected unemployment in the early 1970s.
"Didn't you feel bad mooching off the government?" Brian asks,
and I can't help but give him the Communist Party line. "You
know the government is just a rich-men's club of fat-cat business
men who stack everything so it tilts their way." Brian laughs, but
I can tell he's unsure how much of what I say I truly believe. And
so too, maybe, am I. I certainly wouldn't mind going back on unemployment for a while. Maybe we will get laid off.

After lunch, we finish up the last of what we can do. The rain
has stopped. At 1:15, we pack up the tools and head off to the
other site. I take one last pee in the Sani-pot and that's it. We
work the rest of the afternoon at the Blair remodel and when I
ask Vern about tomorrow, he says to report back to the Blairs'.
"This is your new home," Bud says, as if to confirm the fact that
we will not be returning to the Chin house, that our part of that
work is done.

What remains for the Chin house? The rest of the sheetrock
to be hung, taping and mudding, painting, hardwood floors and
carpets, tile, final plumbing, final electric, doors, trim, case and

base, landscaping, the heating hardware, the security and phone hookups, the garage doors, the kitchen cabinets and appliances, and the outside patios, decks, driveway, and landscaping—all of this, I think, will be done either by Vern and Bud or by sub-contractors. As of now—I imagine—Brian and I have moved on to the next one. Sitting in the diningroom today, I allow myself, just for a moment, to go back to the first day when I stood, cold and bored, with my eye to the leveling instrument, looking down at the plans, trying to imagine what was to emerge. I remember the exact spot where we now sit, remember it covered with grass, remember the blade of Spike's bulldozer digging into the first cuts of black earth, and all the steps in between. In my mind's eye, it's a fast-frame movie, over almost as soon as it has begun. All the long, drawn-out time torture of those dark hours of melancholy, boredom, and despair, now like the dust on the floor of the diningroom, blown away with the broom, forgotten, buried in the walls of the house. We leave no offering, no sacrifice, no can of chili in the framing, only a few doodles here and there, numbers written on wood, bent nails, a little blood, a scrape of fingernail, the waffle-face impression of a hammer, memories, our own memories of what we put into it. Here is the house, still six weeks away from occupancy, but fully formed, al-ready alive; as the carpenters move out, the carpenter ants move in—Brian found a few two days ago, and asked me—"Should I tell Beverly?" A wasp was building a nest up under the eaves yes-terday, there are traces of birdshit on the roof. The golfers will be bombing away with their balls from the third tee. As we pack up our tools and back away I even forget to take one last look. "We're outta here," is all we say. And we're gone.

May 2

B UT NOW, a month later, we're back. I thought, from what
Vern said, that we wouldn't be doing any of the finish work
on the Chin house, but now it turns out that we will, at least
some of the crude finish work. It's a Tuesday, and Vern has left
me hanging since last Thursday, when we finished up the siding
over at the Blair remodel. We've been working mostly at the
Blairs', with one small side-job redoing a deck on a big house
near the country club. In a month, the Chin house has been
completely sheetrocked, taped, mudded, textured, and painted
on the inside. In fact, the painter is still inside today, spraying
the last coat on the upstairs.

Vern calls at about 8:30 to tell me to come on over. I cop a low-
level resentment about this on-again, off-again scheduling, but
it seems to be in the nature of construction work, and on the re-
alistic side I can imagine how hard it is to keep work lined up for
me and Brian one hundred percent of the time. When I worked
for Paul Murphy in Santa Monica, we would sometimes have
down days where we washed the company trucks and rebuilt
storage facilities in the office, but Vern does not hand out work
gratuitously. And I feel it gives me a little righteous edge in an
odd way. So, since he only called this morning, even though I was

certainly expecting his call, I dawdle and indulge myself, drag my feet, chat with Kathleen, goof off with Shannon, let the time slip by a little, and don't show up at the site until about 9:30.

Vern and Bud are already on site and so is a delivery truck with all the doors and finish wood. I help them unload it into the garage. It looks as if Vern and Bud were probably here on Monday and maybe even on Friday, doing some of the last few jobs on the outside. Vern has built an intricate, removable cedar deck onto the upstairs patio, and Bud is chipping away at the last few, complicated pieces of siding around the multifaced front entranceway, which now has its two glass-block sidelights. Brian is not here yet.

Vern sends me into the house to start sweeping out the studio and back hallway, which will get a half-inch particleboard underlayment in preparation for a parquet floor. I guess the rest of the house is carpet. I don't know what the kitchen is getting. The painter is spraying and the house is completely closed in. The heaters are on and there's a heavy toxic smell of latex throughout. I start sweeping, and Brian shows up about 10:30. We get right into it. Vern and Bud are starting in on the finish work. Building window surrounds, hanging doors, finishing out the closets. The house is dark in most areas because the windows have been masked out with plastic, and the plastic is coated with overspray from the painting. We cut our particleboard out in the garage and drag the pieces in. Vern has set the tolerances fairly tight, a quarter-inch where the boards meet the wall and a sixpenny nail's–width between sheets, so it takes some time to get it right. We have to lay building paper down as a vapor barrier as we go. The air is warm and humid from the heaters and the paint smell, and we are soon down to tee shirts, even though outside it's sprinkling and cool and overcast. The day goes along smoothly, quicker than usual because of the late start. Brian seems tolerable, maybe because I haven't been around him in almost five days. Once the underlayment is tacked down, I go over it all with the nail-gun, nailing it with eightpenny ring-shanks every four inches along the edges and every six inches in the field. The gun is deafening in the stark studio space and I wear ear protection, which cuts me off completely from all small talk—a real blessing. The work is smooth and not too taxing physically. I wear my foam knee-pads and have new boots I bought at G. I. Joe's on Friday when we were off work. They're cheapo leather boots, just like the ones Vern buys, except they only have the six-inch top, not the eight-inch like his. Vern, I notice, has finally broken down and bought himself a new tool belt. Hard to believe he's finally hung up the old one—it looked like he'd been using it since sometime before Nixon became President.

May 3

Another day at the Chin house. We start out doing the last of the underlayment in the laundry room and then upstairs in the guest bathroom.

After lunch, Vern puts us to work on the window surrounds. These are clear fir boxes that need to be made to fit exactly the dimensions of the vinyl windows that were installed from the outside before we started laying up the siding. The depth of the windows varies slightly, so each one needs to be custom ripped to width on the table saw, four and five-sixteenths, four and three-eighths, four pieces of wood per window. As usual, there is almost no extra stock for fuck-ups. We are fabricating the surrounds and Bud is coming behind us and nailing them into place. Brian and I fall into a separation of tasks. I figure out the measurements and give them to him, he rips the wood down and cuts it to length, I sand it and nail it together. We start into a pretty strong rhythm, and because we are working in different locations, there is not much talk. The first part of the afternoon runs smoothly along.

About 2:30, I look out the window of the dining room and see Bud and Vern and Brian all walking around the front of Vern's truck. They walk around the truck and then they gather around the passenger-side door. I can't figure out what they're doing, but I assume they're looking at the plans, or perhaps looking at some small job that needs to be done on the south-facing garage wall. I'm curious, but I go back to nailing the ripped fir boxes together and only go outside when I run short of material.

When I do go outside, Vern and Bud are still standing around the truck and Brian is sitting on the ground with the middle finger of his right hand in a huge, bloody bandage. "What . . . ?" "He cut himself on the table saw," Bud says. "Oh shit," I say. I can feel it go right through my body.

In fact, I had been watching the way Brian was cutting on the saw earlier, and I almost said something about it. He was pushing the boards through with his left hand on the back end and his right hand right down on the board, with the palm passing only a few inches above the whirling blade. When I use the table saw, I never even put my hand on the same plane as the blade if I can avoid it. I always push with my hands from behind until the end, and I usually use some kind of push stick to get the last part of the board through the gate. But I didn't say anything, because—as I think later—I see Brian do so many dangerous things, and he obviously resents being told anything, particularly around safety issues—well, I guess I just stopped telling him things. And now he's almost cut his finger right off, the middle finger of his right hand.

"How bad is it?" I ask. "I don't think I got the bone," he says, with a kind of sick smile on his face. And then he holds it up. It's the classic middle-finger salute, now exaggerated with that bandage. "I won't have any trouble flipping people off for a while," he says, joking. "I'm so sorry," I say. "How's it feeling?" "It's starting to hurt pretty good," he says. Vern is calling around, and finally sends Brian off with Bud to St. Joseph's to get the thing stitched up.

When they leave, Vern seems totally pissed off. It will surely be a bad spot on his safety record. And he comments too, that the table saw is the one saw on the job that has no safety guard. Even though they come equipped with safety guards, table saws are almost impossible to use with the guard in place. The guard binds on the wood in many cases and the operator ends up taking bigger risks trying to free up the wood than he would take just running it straight through an unguarded blade. I have never seen a table saw with the guard in place. When a contractor gets a new table saw, the first thing to come off is the guard. Still, even when I tell Vern all this, all information he surely knows already, he remains grim and angry.

We go back to our respective tasks. Now it's me pushing the wood through the table saw. There is blood everywhere, on the saw table, on the stack of wood beside the saw table, in the sawdust at my feet. I work ahead methodically, leaving the blood spots facing out on the surrounds, where they will be hidden against the framing after the windows are cased.

For the rest of the day, I have an almost sick feeling of anxiety and apprehension, and whenever I rip down a piece of wood on the saw I try to remain as focused and present and alert as possible. We work around these whirling, deadly sharp blades all the time—the Skilsaw, the table saw, the chop-saw, the big, old, radial arm saw with its ten-inch blade, and we tend to take them for granted, to forget how quickly they can cut into you. I have heard stories of carpenters reaching down and picking their fingers out of the sawdust pile at their feet, driving into town with the things in ice, to get them stitched back on. And I consider myself lucky to still have all my fingers intact. I feel terrible for Brian. Nineteen years old. That middle finger will never work quite right again. As always, when something like this happens on a job, my first and deepest and most powerful reaction is— thank God it wasn't me.

May 4

No Brian today. I come in and find Vern in his truck sipping on a thermos cup of coffee. "What's the story?" I ask, and he says, "What story?" almost playing dumb. "About Brian," I say. It seems so obvious. "What was the report?" "Oh," Vern says, with that look of weary resignation on his face. "He cut a tendon, he'll be out for at least three weeks, hand in a cast." He says this as if he's totally pissed off at Brian for being so stupid. There's very little compassion in his voice. It's almost chilling, as if the thing he's most worried about is his worker's comp rates going up. I don't ask for any more details, and we start to roll out the tools.

While we're carrying the long chop-saw table out of the garage, I step on a nail. It's the first time in a long time, maybe since Malibu, that I've stepped so directly on a nail. This is an eightpenny galvanized nail that is sticking out of a piece of plywood and it goes directly through my boot sole and deep into the center of my foot. It hurts like a mother-fucker. I drop the chop-saw table and I'm dancing around trying to pull the thing out. Thank God I have the new boots on with their fairly thick soles. In my old work sneakers, the nail would have almost come out the top of my foot. Is there a Christ metaphor in this somewhere? Or am I simply injuring myself in a sympathetic reaction to Brian (or is it survivor guilt)? Anyway, I hop around for a while.

Vern finds the offending stick of wood, pounds the nail over, and throws the thing angrily into the scrap pile. It's been a rule on this job, as on most jobs, that all nails sticking up out of scraps of wood be pounded over, so I'm not sure why that nail was there. But I'm half suspecting it was something Brian did.

Still, I say nothing except to reassure Vern that I've had a tetanus shot in the last five or six years. And that I'll just put a Band-Aid on the cut and hobble around for the rest of the day. I'm not about to walk off the job and file a worker's comp claim. The puncture continues to hurt for a few hours and then slowly calms down.

Later on, both Vern and Bud cut themselves slightly, and a water line under the studio is discovered to have been punctured by a nail while I was nailing off the underlayment (and this is not, I'm quick to perceive, my fault, as all water lines have to have metal plates protecting them from nails—plumber's job). By 9:30, all three of us, plus, in a way, the house itself, are bleeding. I point this out as a kind of karmic group psychosis, but Vern and Bud just look at me like I'm a bit weirder than usual.

The pace of the work picks up. I'm back to building the last of the window surrounds and then, when those are finished, installing them. Vern, it seems, is slowly and grudgingly giving me more responsibility. The placement of the surrounds is demanding, since they need to be equally spaced against the window edge on all four sides, and straight and square within a sixteenth of an inch. I do the big diningroom wall, which consists of eight windows that all have to line up with each other as well. It's a joy to be working alone and in silence. The day goes by quickly and easily, and with very little strain physically, as the wood and nails are all small and lightweight.

Toward the end of the day, Brian shows up with Jennifer. He now has an enormous bandage on his middle finger, though his hand is not yet in a cast. Apparently he cut the tendon on the top of the finger, so his gripping tendon is still intact. He goes over to the table saw and tries to demonstrate exactly what he was doing when the accident happened, but we all—Brian included—have trouble imagining how he could have cut the outside of the finger. He would have had to have the hand turned upside down. "Maybe you were briefly abducted by aliens," I say. "That's a possibility." He describes how they screwed a screw down into the center of the finger to keep it rigid and how at first they were screwing through a part of the finger that was not quite numb from the Novocain. Everyone winces. It sounds like bamboo shoots under the fingernails. They'll be taking the screw out after a week and then putting the hand into a cast. Jennifer gets a copy of the worker's comp claim form and Brian, though he has been told he can come back for "light duty" next week, acknowledges that he will really not be able to do anything since it's his right hand, the hand of the hammer and the saw. He says he figures he'll just stay on worker's comp until the hand is functioning again. As bad as I feel for him, I can't help the feeling of relief that I won't be working around him for at least three weeks. A vacation from Brian!

The day wraps up easily and I'm home by 5:00.

May 5

Vern is taking his whole family to Disneyland, starting today at 4:00. Two cars, the trailer, he and Connie, the kids, their spouses, and the grandkids. They will be camping at a KOA campground just a few blocks from the Magic Kingdom and will be gone nine days. We come in this morning to do a few last-minute pick-up jobs at the Chin house, and then clean up. Vern has me scheduled to do the window surrounds and the underlayment at Blairs', but that will only be for two days or so.

I have lined up a three-day job in the middle of the week, just by lucky coincidence. My old friend Lou has hired me to play "Joey, the mechanic" in a National Health Institute video about problems families have dealing with aging parents. Bizarre. I only tell Vern and Bud that I have another job for three days and leave it at that. They think I'm strange enough without having to know that I have a side job as an actor.

By noon, we have moved all the tools I will need over to the Blairs', and I spend the rest of the afternoon working there alone. I bring my radio from home and listen to the music I want to listen to. Rush's obnoxious voice is silenced for now. No one talks to me. I can stay completely concentrated on what I am doing. Time flies and the work moves ahead smoothly. We will not be back at the Chin house until mid-month.

May 16

TUESDAY. We make it back to the Chin house about 2:30 in the afternoon, after a final six hours at the Blairs,' where we basically finish all the case and base and clean out the last of the tools and materials. Brian shows up about 1:30. He comes up behind me, stealthy and silent, and I suddenly look around and there he is. Huge! I forget how big he is when I haven't seen him for a while. He's all smiles and says everything is going fine. His finger is not in a cast but is hanging right out there in front of us, a finger the size of a medium zucchini with a huge, ugly scar bristling with black stitches. To make it worse, the shiny, stainless end of the screw they drove down into the center of the finger, a screw that keeps the thing rigid while the tendon grows back together, is just visible, sticking out of the tip of the finger. Brian shrugs, "It doesn't bother me, I don't feel it at all." But just the sight of the thing makes my skin crawl. Someone was telling me recently that your hands are so full of nerves and pressure points, so uncommonly sensitive, like your tongue or your earlobe, that when you cut anything on a hand, it has an impact way beyond what a similar cut would have on another part, a leg or an arm, say. We have some brief small talk with Brian, and then it's on about our business, closing down the Blair job and getting on to Chin's. Brian has just stopped by to visit. He is still a few weeks away from returning to work.

When I see it again, the Chin house fills me with despair. The fake Mediterranean tile roof, for one thing, looks so cheesy and cheap, I can hardly bear to look at it. A piece of tattered Tyvek hangs by the front door, where a few last runs of siding are still lacking. The midday sun is bright, almost white, and the place looks bleached-out, dusty, and uninviting, like some half-finished, abandoned, repossessed mini-mall in San Bernadino. A pile of debris has grown beside one of the garage doors— old bathroom-fixture cardboard, paint-splattered plastic, coffee cups, bits of wood and metal. Inside, the painters are still working. They have lacquered the cabinets, which look far better than anything else in the house, dark wood, genuine, expensive, but they are emitting a vaporous, toxic stench, a fume cloud that seems to go straight into the center of my brain, where it begins work knocking out circuits and triggering nausea reflexes.

We're here to do case and base. Case is the casing, or "trim," that runs around all the windows and doors. And base is the baseboard, which is laid in everywhere where the walls meet the floor. The case and base at Blairs' took about a day and a half. This place is a whole different story.

We go into the studio room, me and Bud and Vern, and look at this humongous heap of sanded and primed case stock—it's overwhelming. I feel small and weak and defeated. Like I can't go on. But this is real work in the real world, and as always there's no real choice. Vern sighs and waves his loose wrist in the general direction of the pile. "Start slappin' her up, I guess." He has no old master's insights into the process, and so Bud and I begin. We set the saw table up outside in the patio area just behind the diningroom. I go from room to room and start measuring every window, numbering them, making a list. For windows, the case is cut a half-inch bigger than the inside dimension of the window surrounds. Each side is cut at a forty-five-degree angle. In the end, when the four sides are laid up, you have what should be a perfect frame with four perfectly joined, mitered corners and a quarter-inch reveal of the edge of the surround board on all four sides. The house has thirty-eight windows but we are only going to case thirty-six. The two out in the garage are not getting anything at this point. I start in the guest bedroom to the right of the front door and work my way around in a big circle, measuring, writing down numbers, giving them to Bud. We have only two hours before we knock off and we don't get too far. The thought of this process fills me with a sensation of terminal exhaustion.

May 17

I have completed the lists of all the windows and their dimensions and Bud and I make a decision about the division of labor. I tell him I will cut them all and he can start laying them up. I'm not sure why I do this, but I think it's because I'm deferring to his skill—one of his true strengths as a carpenter seems to be finish work. He's done a lot of this case and base stuff. I watched him at Jenny's. He's good. Plus, I want to do the cutting, to make myself a little factory in the diningroom with the chop-saw and the chop-saw table and my endless pile of stock, and start cutting these babies. Thirty-six windows means I have to cut 144 pieces of casing. The degree of accuracy has to be within about a thirty-second of an inch. Vern starts the day off, for example, by handing me three well-sharpened number 2 regular Monarch writing pencils, the kind of pencils you use for school work, the kind of pencils accountants and math teachers use. He makes his point. You can't be accurate to a thirty-second of an inch if you're scribing your marks with a stubby carpenter's pencil that leaves a line an eighth of an inch thick.

Now I am marking, cutting, labeling, carrying the sticks around the house in bundles of four. I'm a chop-saw technician. On some very profound level, I am gleeful to have this job all to myself. No Brian to share it, argue with, battle with over every infinitesimal, nit-picky thing we do. This time around, I make every decision, and I don't even have to say a word.

Bud, like me, goes off into the corners of the job site and communes with himself. Vern is the same. These are not highly gregarious, highly socialized creatures here. Bud is starting to lay up the case in the kitchen and he brings out the radio and has Rush Limbaugh fulminating away well within my earshot. But the radio has some sort of faulty wire, so the station comes and goes. Not a wavering signal, but an on-again, off-again signal. You never quite know when that nerve-grating voice is going to jump out at you from the corner, mouthing off in some apoplexy against Clinton and Democrats and liberals and tree huggers and feminazis and all that bitter juice he spews.

Mostly, I am in my own world, though. The saw whines and cuts with a quick, decisive thunk. I work my way through the list of windows and dimensions. And I am getting highly accurate. I have a mark on the underside of the saw table, like a gun sight. I squat down with my glasses on and sight down into the dim shadows under the wood to find my mark, then I line it up with the thin pencil line on the caseboard I'm cutting. Once it's there, I don't even have to look. Saw on and down, chop, one quick cut, and the board is perfect. Most of my cuts now split the pencil line.

I try to keep the work area clean and to keep the little butt ends out from under my feet. I do not cut a single cut without considering the wicked power of that spinning blade and what it could do to my fingers. This saw has an orange plastic guard that flops down around the blade. But still . . . I'm reaching in after every cut, clearing away the little chopped triangles that are left over from these forty-five-degree corners. Everything is coming out sweet.

I'm done with all the cutting by about 1:00, and so Vern has me start in with the nailing up. He seems to be operating with a lower tolerance for error on this job and has us gluing all the corners as we set them. We're only going to nail the inside edges up, then glue the corners, nail the corners and then let the glue set up before we nail the outside edges, and that will be with bigger nails, the nails that really hold the case to the walls. But that will be tomorrow and with the gun.

Today, the whole problem is corners. Some of them don't fit quite right and you have to take the small, palm-sized block plane and shave paper-thin curlicues off one end or the other, then press the cuts together again, check it for "gaposis," squeeze it, play with it, finally accept it. Say to yourself: This is the fucking best I can fucking do and I'm gonna fucking leave it at that. "Ain't no pianer bench . . . can't see it from the freeway . . . close 'nough for government work . . . can't see it from my house." Glue it up, nail it up, move on.

I spend the last three and a half hours working at this and only case out six windows. Vern has done thirteen over the last day and a half and Bud has done seventeen. Of the three of us, Bud's work is the best—the straightest lines, the tightest joints, the most even and consistent reveals. Vern is fast but not quite as clean. Still, he did a wall of eight windows that is the centerpiece of the dining room and they look pretty god-damned perfect. The lines of the casing of all the adjoining windows line up both vertically and horizontally, and I know that shit is not easy. I go upstairs into the small bedrooms and do the windows there one at a time. None of them are side by side. My most perfect window is the one in the small guest bedroom above the diningroom. I'm proud of it, but I don't show it to anybody. I don't say a word. Nobody says anything to anybody about their work.

Silence on everything but fuck-ups. Speed and perfection are the standards and not worth mentioning.

May 18

A beautiful, clear, sunny day with blue skies and warm breezes. But we are working inside. The whole timetable for the house has been in perfect reverse. Dig it out, frame it, roof it, and side it while the rain pours, the winds blow, and the temperature creeps around just above freezing. Then, when the warm, balmy days of spring start in, when the flowers are in full bloom, and the grass is a lush, bright green, go inside where the air is poisoned with the remnant traces of sheetrock mud, insulation, plywood glue, paint, and lacquer, and work all day like a troll in a cave. But that's the story here at the Chin house as we push toward completion.

And today I am on door casing all day. I case out twelve doors, not all of them on both sides. In the end, I cut, build, glue, and nail twenty casing sets, sixty pieces of wood total. I only fuck up two cuts and they are both on the same door and they are both at the end of the day. I was up playing hockey last night until 10:30, got to bed after 11:30, did not sleep well, woke up about 2:30, read a book about the deadly Ebola virus for an hour, went back to sleep, then got up at 5:30 for good. I'm a somnambulistic pre-coma case when I drive off to work at 7:45, and by the end of the day I'm toast. My body feels tight and crimped over. I'm up and down the ladder, up and down the ladder, and for at least half the day I'm up and down the stairs to the second floor and then up and down the ladder.

Bud is doing base and Vern is picking up the odds and ends, risers and treads for a while, and then cap pieces for the stair surrounds, custom trim for the pocket doors, and so on. We are almost completely silent with each other. Even at break and at lunch there is nothing for us to say. For a while there's a discussion with the backhoe guy about the various trucks Ford offered for sale in 1978, and I'm sitting there like, "Was the Lariat the biggest truck that year or was it the XLC?" as if I know anything, or even more absurd, as if I care at all.

Sometimes, like the first half of lunch today, while I am sitting alone with Vern in the sun and I have asked him a couple of leading questions about some as yet unfinished corner of the house, and he has answered in as brief and monosyllabic a way as he knows how, we will sit for long minutes in absolute silence, munching on our sandwiches and apples, and saying not a single word to each other. I'm waiting for him to say something, ask me something about myself, maybe, something about my past, my life, my family, even about my car, which is an old Honda stationwagon, and, I assume, beneath contempt—but no, he says not a word, and we are stuck in this long awkwardness with no conversation at all. And then I think, well he must like it like this. Isn't this the way of Gary Cooper? When Bud comes back, we talk for a while about the new cell phone he's going to buy, and then the backhoe guy sits down in a pile of dirt and talks about his old Ford trucks.

When it's back to work, it's back to work in silence. Measure the top piece, cut it, nail it up, quarter-inch reveal on each side, measure the two sides, cut them, bring them back to the door, check them out. If they don't fit perfectly, get out the little pocket plane and shave them down, a sixteenth, a thirty-second, a cunt hair, a red cunt hair, until they fit perfectly, then glue, nail, set, finish nail, move the ladder, bend over, nail the bottoms, and then on to the next door where the process starts again. Measure the top, add a half-inch, cut it on the chop-saw, and so on.

My last doors at the end of a long day do not go well. I am struggling, trying to keep myself in the game. I say my mantra, I start the Lord's Prayer a hundred times and never get past "for-

give us our trespasses." I sing "Amazing Grace" under my breath,
I whistle, "There are roads that run through summer sun/
Through fields of waving corn/And streets that pass through
cities broken, mournful, and forlorn . . ."

I seem to miss Brian and it's hard for me to admit that to myself. Sure he's loud, obnoxious, bossy, and a know-it-all control
freak. But he's light-hearted too, he has fun, he likes to joke
around. I wonder about the timing of his accident, the thing I
had once wished for so heartily; Vern said he couldn't imagine
having Brian do the finish work on the Chin house and then exactly on the morning of the second day of the finish work he cuts
his finger on the table saw and will be out of commission now for
the duration. Certainly we will be done with this, done even with
this monster pile of case and base, by the time Brian gets the
green light from his doctor and straps his carpenter's belt back
on. Funny as I think about it now—my whole conflict with
Brian. For one thing, I'm almost sure he was never aware of it. He
seems to regard me as his true friend, and has called me several
times in the past few weeks, just to chat and gossip about the
job. Really, the whole thing was such a symptom of my fucked-
up thinking; my pride was so wounded by having to work as an
equal with this kid who, as I saw it, knew next to nothing about
carpentry. Wasn't I the old veteran, the seasoned professional?
And yet the hard truth is this: In many ways, he was a better
worker than me, real working-class for one thing—work is in his
bones. Plus, he's faster in some places, more productive, more
daring. Sure, he fucked up a good deal, but in the final measure
of our relative worth, in Vern's eyes at least, bottom line, we
might have come out about equal. Well—whatever. I do miss
him now, miss him as my only ally on the labor force, miss him
as a companion in rank, and, yes, indeed, miss him as a friend.

I finish up the day in despair and drag away home to a quick
meal and softball practice with Shannon's team. At the end of

the evening, I talk to Greg on the phone and tell him about the silent miseries of my work and he asks me why I stay there, why I keep doing it, and I tell him, "I'm staying there because it's such a challenge, because I believe ultimately that all my happiness is there, that if I can find perfect happiness on this job site, in the very middle of all this sorry, sullen drudgery, then I can move on, move on to the next level, whatever that may be." And he says, "That's amazing because it's like you're a monk. That's exactly what they do to monks when they start out. They give them eight hours of totally mindless, boring, repetitive labor, labor with no end in sight. And that's how you are. You get to be a monk for eight hours and then go home to your wife and kids, your family. You get it both ways." And I say, "That's right. I am a monk. And the funny thing is that being a monk is what I always wanted to be. Only I could never give up all the other stuff." "And now you have it both ways," he says.

We talk briefly about the geodesic dome project, but it seems the client is going to go with another contractor. That may be just as well. I don't think either of us is quite ready to take on such a big project.

May 19

Today I'm a monk and I go to work like a monk. I can't wait to get to work to try out the new theory. I'm a monk. I wonder what Vern would say. Not that he ever asks me. But if he ever did ask me in the morning, "How's she goin' today?" and I said, "I'm a monk, today I'm a monk and nothing's the matter . . ." He'd just shake his head and say, "That right?"

I start in where I left off, with the casing of the last side of the last door. My body feels better. The morning is crisp and clear and inviting, but I am inside, in what will be Beverly Chin's bedroom, and the air is dark and damp and musty. The casing goes together so perfectly. The casing, glue, nails, hammer, chop-saw: these are my holy articles. It all seems so simple.

When I am done with one casing, I help Vern load the last pile of construction debris into the back of his trailer. This is the stuff that bothered me so much two days ago—plastic, old garbage, bits of wire, particleboard, cardboard, even one last survey stake from the moment before we first cut into the ground in November—leftovers of all kinds, but now, outside in the morning sun, I work myself up to a sweat hauling and throwing this stuff away.

After the pile is gone, I work with the nail-gun, nailing off all the window and door casings, none of which have anything more than the minimum number-four finish nails around the edges to hold them in place. Now I have sixes in the gun, and I nail the outer perimeter. Sometimes, where the plaster behind is at uneven levels, the glued joints break apart and leave gaposis. I keep running back to the ones in the front room, the ones that Vern did, checking them to see the standard, and when I see gaposis in Vern's work, I am joyful and relieved. As long as my gaposis is smaller than Vern's gaposis then I am tracking along at a workmanlike pace. Bud's work is crisper, but even his glued joints develop hairline fractures here and there. Later, I talk to the painter and he says that he's seen much worse gaposis on other jobs, and that they can putty anything, and I believe him. I may be a monk, but I'm a perfectionistic and uptight and anxious monk nevertheless.

In the afternoon, Vern and Bud leave to go back and finish up Jenny's place. The painter leaves, the backhoe guy leaves, and I am alone in the house for the last hours of the week. I think this is the only time in the entire project when I have been alone here. I am finished with the nailing and I am laying down baseboard back in Beverly Chin's bedroom, the room where I started out this morning.

There are now rooms in the house that are virtually finished, except for final painting and carpet and drapes and furniture. Today, Vern even said, "It's starting to look like a finished house," which is a real mouthful coming from Vern. But it's true. And now, too, I can look out the window and see that the grade is flattened out to its final level, the gravel, where the concrete driveway will be poured next week, is in place, the debris is gone, the golfers are going by, the end is in sight. What's left? A few doors, the closets, the last scraps of siding, some of the banister treatments, the tile and the parquet and wood floors, the carpet, all the final sanding and painting, the final plumbing fixtures, the sockets and switches and lighting fixtures, but most of that will be done by the subs. I'm not sure—there always seems to be more work than I think. But I can't see more than two or three or four more days of carpentry. Next week may be it.

I'm reluctant even to ask Vern where we go from here. I'm afraid he has a big house looming directly on the horizon. When we are alone again for lunch, and silent, I think of asking him, "So what's up next on the agenda?" But I don't. I'm afraid he'll say he has another big house to start. And also, I want so much for him to say, "Well, it looks like we'll have some down time here for a little bit." The funny thing is, I almost don't care either way. If we go right ahead into the next job, I'll be outdoors, back to digging and footings and foundations, with dirt on my hands, but with the sun on my back. I'll be an outdoor monk in the summer. If we start another big house next month, we could be closed in by October. Who knows?

Today I'm happy and absorbed. I work and pray and believe in the sanctifying nature of endless, repetitive tasks. It's Friday, so I guess I'm free to believe in anything. I've worked a full forty this week, and in the end, driving home, I stop at 7-11 and buy a Pepsi and some Hershey's kisses and a couple of little mini-Nestlé Crunch bars to celebrate.

May 22

ONDAY morning. I get to work late. No matter how deep
my resolution is to arrive on time, sometimes I seem to
give over those last few crucial minutes and leave, like today, at
five of 8:00 for my seventeen-minute commute. I make a resolution, as I drive across the bridge, to arrive before Vern tomorrow.
Today he is already cutting boards when I get there, cutting
lengths of clear fir for the painters to sand before we case out the
last doors in the studio end of the house.

Finish work. I can't resist the constant feeling that we are
finishing up here and that we won't have work for a while, that
we will be between jobs, that I will be laid off, that I will break
away for good from Marshall Construction Company. The
thought fills me with exhilaration. And again, Vern says nothing
about what we might or might not be doing next. And again, I
don't ask.

I spend the morning doing the rest of the baseboard in
Beverly's master-bedroom wing. The bedroom, the sink area,
and a huge walk-in closet with eleven different wall surfaces. It's
hard to imagine how this is possible, but it is indeed true. The
closet is built around part of the back side of the guest bathroom, and includes a dead zone where all the heating ducts

chase down from the attic area where the forced air unit is located. By now, I'm almost a master at this baseboard routine. The work involves a few tricky cuts and I'm up to them. I'm finding studs with an electronic stud sensor that lights up a row of red lights when it passes by a stud. Like sonar. The work goes well. I'm alone, I'm rested after a weekend of completely goofing off. I actually sat in the sun for a few hours and read a LeCarré novel, a major break from my compulsive routines.

Another day without Brian, too, and that seems another blessing. The time doesn't actually go any faster without him around. But I remain in a much more even emotional state. No challenges, no conflicts, no arguments, no slave mentality. The whole thing is now me and the work, and I work steadily and carefully and well.

No comments from Vern. No approval. No positive reinforcement. I actually get called on a couple of things. In one instance, Vern tells me that some of the casings around the big windows in the studio have not properly seated against the wall. Late in the day, I get a little grief from Bud and the painter about the number of nail holes I left in the baseboard along one wall of the master bedroom. No big deal. The worst thing that happens all day is in the middle of the afternoon, when one of the painters takes a shit in the bathroom right next door to where I have to put up a one-by-four ledger for a closet shelf. He is even apologetic as he goes in, and tells me, "I'll try not to stink up the place too bad." The Sani-pot is gone, another milestone as we move toward closure.

A few days ago, I asked Vern when he expects to have Beverly moved in, and he said probably by the middle of June. And meanwhile, I can't see much more for us to do. The hardwood-floor guys are coming in at the end of the week; we will have to wait for them to finish before we can do the last of the case and base in the kitchen, diningroom, and livingroom areas. All I see left are a few closets and the siding on half of the small wall that surrounds the front door. A post deal of some kind is supposed to go in at the corner of the overhang over the main entrance, but that will definitely be a job for Vern.

Finish work—as in finished. I can't wait to get out of here.

May 23

Another day of finish work—just me and Bud and Vern with the painters in the background. The day feels routine and easy. I start out in the front hall closet, nailing up one-by-fours and then fitting an irregular, slightly complex shelf into place against the odd-angled walls that have been created by the curving bay front of the entrance. From the front closet, I go to a utility closet under the stairs, a room just off the garage and the laundry room, where I have to build a series of shelves into the two corners, using the heating duct to support one leg of each shelf. The job is not easy, it requires a certain level of ingenuity and patience, and I click along like an expert.

From time to time, I think about what it would be like to try to do this stuff with Brian around. I've told a few people over the past weeks, when they ask me how work has been going, about the fortuitous accident that has removed Brian from my life for this phase of the work. I know how edgy it would be to collaborate with him, particularly if he was in one of his Bad Brian moods, trying to do this fussy little shit.

The fact is, I'm absorbed all day, I feel good, my body moves are easy and pain-free after these weeks of small-dimension lumber, small hammers, and finish nails. The air outside is in the low eighties; if we were out framing or sheeting we'd be dripping sweat, but inside it's cool and comfortable, with light breezes circulating through the open doors and windows. I work in a tee shirt and jeans. My new boots are beginning to feel like cushion air. Not much talk, not much joking, the usual long, silent stretches during break and lunch.

For a while, we talk about different ways to catch the kid who has been vandalizing the site for months, the same kid, we assume, who put the stamped set of plans into the shitter a few months back. His latest number was spray painting the word "FREK" (presumably standing for "FREAK") on a small concrete slab outside the back garage door. He has also recently fucked with the backhoe and cat that were left here overnight, glued all the door knobs with plastic pipe glue, thrown dirt clods through the air ducting, and pulled the wires out of the garage-door openers. The suspect is the kid next door, a wan-looking ten-year-old who can occasionally be seen shooting desultory baskets outside his garage. Neighborhood gossip says he's a troubled kid with a history of low-level misdemeanors. Parents never around, left alone a lot. Plans for confronting him are offered, but none more reductionist than the painter Dave, who proposes that we simply go next door and tell the kid's mom, "Look, if one more fucking thing happens over there at the site, we're just gonna come over here and burn your fucking house down." I propose a stake-out. Bud wants to set up a video camera. Vern, who has filed a few claims with the insurance company, says they would be happy if we could catch him. He figures the total damage is up near a couple of thousand. Deep down, I don't give a shit. It ain't my fuckin' problem—and thank God.

Mainly I'm eager in my work, happy to have it, happy each time I walk back into the diningroom to cut another piece on the chop-saw. Bud is out in front laying up the last few scraps of siding around the front door, Vern is installing an elegant, clear fir banister rail on the stairway, and I am back in my closet with my shelves. The painters have been masking off all the windows and will start spraying tomorrow, so I am half anticipating a day off. I know we have done almost all the work we can do until the floor people come in to lay the oak floor.

As usual, Vern says nothing until the very end of the day, when I am in the bathroom again, this time setting the last of the baseboard. He comes in to change his clothes. He seems to be on his way somewhere and needs to wear more-formal jeans and a striped jersey shirt with a collar. While he's on the toilet taking off his work boots, he tells me we'll be going over to the painter, Frank's, warehouse in a neighboring town tomorrow to put up a partition wall. I'm to meet Bud somewhere along the road and follow him to the site. My heart sinks a little, because I had been hoping for a reprieve, but then I remember, I'm a monk, and I shift from despair to acceptance. What the hell, it'll be fun anyway, to get away from the Chin job again for a few days, to do something different, to get back to a little framing.

Here at the Chin house, there can't be much left. A few last touches on the closets, and then the door casings and baseboard in the kitchen, diningroom, and livingroom after the oak floors are down. And then . . . we're outta here.

May 30

ALMOST a full week since I last worked. I had to stay home for a day to take care of Shannon, who was sick, and then Vern called off work for the rest of the week because the painters were going to be spraying the door and window frames. Monday was Memorial Day; I worked at home on a brick sidewalk. And finally, today, Tuesday, I'm back at the Chin job.

No sign of Brian and no word on when he'll be back. The hardwood-floor guys are downstairs laying in the short block oak. The painters are still spraying upstairs and the toxic fumes from the lacquer drift in evil clouds down to the kitchen, where Bud and I are starting to install the cabinets. The concrete people are out in front, preparing to form up and pour the driveway. One of them, a kid no more than twenty or twenty-one, in cowboy boots and a silver belt buckle the size of a tortilla, seems to be in charge, wielding a full nine-pound sledge with an eighteen-inch cut-off handle and calling everyone "Sir." He wears a shirt promoting COED NAKED BASKETBALL—EVERYBODY ENJOYS A SLAM!

With part of the hardwood floor laid and the window and door casing painted, the house really starts to "look like a house," as one of the painters, Dave, says on his way to the can.

Bud and I fuss with the cabinets, moving them into the kitchen and out of the way of the floor people. They are all wood, stained and varnished and spotlessly perfect. I have to keep reminding myself to be careful, not to bump into the walls. Vern cruises around, rubbing his fingers on the dings and bungs where people have been careless, muttering, "Poor walls, poor walls."

In the afternoon, he puts me back on closets, gives me a fairly complicated project involving more shelves, and this time, partitions and rail caps and clothes bars. I have to rip some eight-foot pieces of three-quarter-inch finish shelving down to eighteen inches. The pieces are heavy and unwieldy, and I have an attack of low self-confidence just as I am beginning to run them through the table saw. Vern helps as they come off the table, holding the weight up so the blade won't bind. Halfway through the second cut, I realize that I forgot to tighten down the gate, and that the width is starting to vary. A major fuck-up! I have basically ruined half the stock and there is no more on site. I'm horrified and humiliated. I want to send in for an immediate brain transplant. Why do these things have to happen when Vern is right there watching? He is obviously fed up. Probably wishes Brian was back. After a brief consultation, we decide to reduce the whole shelf unit to seventeen and a half inches, which will work with the stock I have fucked up.

Once over this glitch (when I come back to work after a long lay-off it's like I've forgotten everything—forgotten that constant vigilance), I get into the closet and start to hum. The pieces fall together like a well-made jig-saw puzzle, and I end the day feeling tired but satisfied, and believing that to some extent I have redeemed myself—that maybe it wasn't as bad a *faux pas* as I thought.

Gratefully too, once again, I am witness to Vern's good, solid, Christian patience, because even as he stands there looking down on my pathetic ineptitude (or natural mistake), he does not stoop to chewing me out or cursing me or firing me or anything. If I had been him and I had been running that board through the saw and I had forgotten to tighten down the gate, I would have been cursing myself as I heard Vern cursing himself later over a piece of pocket door trim—"That's it, chew it all to bits, just chew it up!" And I probably would have kicked the shit out of the poor saw in the bargain.

May 31

Vern puts me in another closet this morning, this one off the master bath and far more complicated, with base to cut, partitions, high and low clothes bars, and shelves to be built into a wall with no studs. But I'm smoking. The weather is gorgeous outside, sunny and hot in the sun. The concrete cowboys are working their asses off out there and I am inside, in the early-morning cool of the house, the nail-guns of the floor guys banging away, the painters' radio blaring out the whiny voice of his most supreme high Rushness, the Pontiff of Palaver, Vern coming and going, listening to Beverly Chin complain about the island cabinet in the kitchen, conferring with the tile man, muttering and puttering as he shepherds this big unwieldy project toward its conclusion. Bud is still working on the kitchen cabinets. I come and go, passing through, out to the garage to the saws, back to my quiet closet, where the work clicks along.

One of the floor men is even older than I am, not as old as the roofer's father, but in his sixties I would say, skinny and wiry, but also not the lead man. The lead man looks a little like a lower bourgeois hippie, but this guy is just an old working guy, and he and I have identical tool buckets. Yellow, probably another five-gallon paint bucket like mine, and the same canvas "Mighty Boss" bucket organizer that I have, the one Brian gave me for Christmas. The guy comes into the closet once to ask me how I like my bucket organizer, and there's something so chummy and corny and simple-minded about the idea of talking bucket organizers with this guy—it almost breaks my heart. He has slicked hair like an old fifties greaser and he eats raw green onions for lunch.

In the afternoon, Bud and Vern go back over to the Blair house for some last pickup, and it's just me and this old guy working along. He finishes up about 3:00 and all the flooring is in. He sweeps up and cleans up, and the downstairs looks so slick, with that new, untreated oak down, the smell of fresh oak sawdust in the air, the concrete guys outside moving gravel, pounding in stakes, sweating and shirtless in the sun, and me inside in the cool house. I've finished my closet and now I'm putting the last casing on the doors where the oak has been laid down. The work is demanding, but I have no distractions, and I hit my mitered corners, one after the other, I fly through that shit. I do ten door casings in a couple of hours, and lock up and leave. At the end of the day I'm completely alone.

June 1

Bud has fucked up the linen-closet shelves and also butchered part of the baseboard upstairs. At the painters' request, I have to go up and rasp it down and fix it. The painters are making jokes to me about Bud's craftsmanship, and in a really sick way, after all the history of this, I am gloating and beaming inside, though I don't join in. Plus, Brian's not around to enjoy it with me.

Bud and Vern are away at the beginning of the day as I nail off the ten casings with the nail-gun. The painter made a remark yesterday about some hammer tracks on the case, and how he has to spackle every one, and so I am careful today. I leave all the nails poking out a bit and then bang them home with a thin nail-set. So perfect. I love finish work. I may be doing the best work of my life.

In the afternoon, after I hang the last of the clothes bars, I start to cut and number the base shoe in the rooms where the oak is down. We can't nail it off yet because they will be coming in next week to sand. But I'm clicking today, working fast, down on the clean oak floor wearing my foam knee pads, scooting along, measuring, cutting custom corners with the coping saw, out to the garage to do the long pieces with the chop-saw.

The concrete crew is here in full force, some six to eight workers, and they're pumping the concrete all day, the back patio, the side patio, the front stairs, an intricate pattern with a two-foot border of plain concrete and inner eight-by-eight squares of exposed and polished aggregate. These guys can really work the bull floats. And work they do, out there sweating and moaning in a brilliant, beating sun, smoking cigarettes, covered with globs of concrete, the trowels and hand floats hanging from their Wrangler pockets. The boss is a big, pot-bellied guy in a wide-brimmed straw hat with a red ribbon—he might be some Concrete Cowboy turned Deadhead. He has that general's knack for marshaling his troops—he laughs, he jokes, he talks nonstop—if ever there was a boss man who looked and acted the exact opposite of Vern, it's got to be this guy.

At the end of the day, one of the concrete workers, a black guy in wire-rimmed glasses who doesn't quite fit the profile, is asking Vern how much the Chin house cost to build. "Right around ninety bucks a square foot," Vern says. "I knew it," the guy says. "They insured my house for sixty-eight. Ain't no way you can build no house for no sixty-eight dollars a square foot." "Yeah," Vern says, in his old Gary Cooper drawl, "it's all goin' up all the time." "Yeah, it's all goin' up," the black guy says, "and that's when the wages got to go up too. Only you guys don't like to hear that."

And I could have hugged him when he said that. I'm such a

weenie, I won't even joke about money, or jab, or make any kind of crack. I've been sitting here for nine months now, pulling in my measly $10 an hour and imagining that everybody else is making twice as much. A lot of money floating around. The Chin house must be coming in right around $250,000. Plus, with the golf course lot, probably close to $300,000 by the time she sets her foot in the door of the finished product. We're getting close.

June 2

A Friday. I'm on a weight-loss kick, so I'm up at 5:00 and off riding my bicycle for an hour. It's the influence of all those skinny, muscular concrete guys with their shirts off—it makes me totally self-conscious about my fifty-one-year-old body. I get to work feeling slightly achy and depleted. The day is hot and the first sections of exposed aggregate concrete are being washed down with muriatic acid. Also the lingering vapors from all the painting seem to cling to the inside air. It feels stuffy and dusty as I work on my knees, measuring and cutting the last pieces of baseboard for the studio, back hall, and laundry room. This area will be getting a parquet floor later in the week.

I'm working with Bud now, and we work in harmony. I even go so far as to tell him, jokingly, to "get the fuck out of my way" when he is using one end of the cutting table to cope out a corner of a piece of baseboard, while I'm waiting to cut another on the chop-saw. And he says "fuck you" back to me in that time-tested way co-workers have of gently riding each other as the work proceeds. This is closer than we have ever been.

Later, as we are numbering and bundling up all the pieces of baseboard, to be stored until the parquet is down, he asks if I have heard anything from Brian. "Not a word," I say. "His baby has got to be about due. Has he called your Dad?" "Nope," Bud says, "I wonder if he's ditched us?" "How long was he supposed to be out?" "Three weeks, I thought, and it's been what, four, almost five?" We don't carry it any further. Perhaps there's too much wishing involved, or even superstition, that if we say it too loud we'll suddenly see his big, loony face appear around the corner with his five hammers, his giant, banana fingers, his mouth going off like an AK-47, his trusty red pencil at the ready.

Vern and Beverly Chin have been discussing the erroneous island counter in the center of the kitchen. Beverly seems adamant that the thing be changed. Vern goes off to hang another door. The rejected countertop is down on the floor, approximately where it would be if the counter were in place. I am putting a piece of finished oak trim into a space between the dishwasher and the oven, and Beverly is down on her knees with twine and string and brown paper, trying to make a pattern for the rounded counter top she wants. I come over to help her, holding the string steady around a small nail while she scribes the arc with a red ballpoint pen held in a loop of the string. This, too, seems far more intimate than we have ever been. And in the process, she confides to me that she can't wait to get out of her parents' house. "Nine months too long," she says with traces of the Chinese accent that has been the source of occasional racist ripples throughout the job. "Well," I say, "it'll be so wonderful for you to get into your own place. And it's so beautiful." "You like it?" she asks, seemingly pleased that I would have any aesthetic feelings. And I tell her again how much I do like it, which is only partly true.

Now that the house is almost finished, and I have a clear picture of what it may finally look like, furnished and occupied, as it surely will be within the month, I can tell that I like the inside far more than the outside. With its towering white walls, bright, bold, white trim, many windows, oak floors, oak cabinets, and the green of the golf course outside, it really is a pleasing and dramatic and above all simple look. The outside still bothers me, with it's fake siding and fake roof, but now that the concrete is going down for the driveway and the mess is cleaned up and the lines of the landscaping are beginning to emerge, I can see that it will not be nearly as offensive as I thought.

We spend the rest of the day doing last-minute pickup jobs. I case out a few more doors, I put in the poles and brackets for the closets upstairs, I go from room to room, trying to see what, if anything, has to be done. By 3:30, I'm feeling vaguely sick, headachy, feverish, and "strengthless of limb," the symptom we found described recently on the label of a Chinese herbal remedy. The day comes to its merciful end for us a half-hour early, even as the concrete squad waits for one last truckload. "I'd hate to be staring at that much concrete at 4:00 on a Friday afternoon," Vern says to one of the workers who has been taking a break before the last big push. Pouring at 4:00, they'll be lucky to get the thing floated and troweled off by 7:00. Warned of the vandalizing kid, they also have a hired guard who will stay and watch until 11:00, by which time the stuff will have gone off and be too hard to mess with.

Vern says he'll call me Sunday night to see about Monday, and a flicker of hope passes through me, that we will have a few days off at the beginning of next week. On the drive home, I feel exhausted and drained and even sicker. After a shower, I climb into bed and try to sleep. I'm supposed to go to a pot-luck for Kathleen's clinic, but I beg off, stay home, and watch the Pacers walk all over the Magic. Kathleen comes home late with an old Frank Capra video, *You Can't Take it With You*. All about the unimportance of money.

Postscript

Eleven months have passed since I laid my last nail into the Chin house. That last day, the concrete crew was outside pouring the slabs for the driveway, I was cutting and setting the remaining pieces of case and base and also starting to feel the first aches of a flu coming on; I had no idea it was my last day and so I didn't even bother to take a last look at the house as I drove away. Just said, "See ya" and "Have a great weekend" to everyone, and headed home as I always did on Friday nights, but without the usual spike of elation because of the sickness starting to come on. Vern didn't call for a week, but it was a week when I was too sick to work anyway, so I didn't bother much about it. I just lay in bed and felt like shit and waited for his call that never came. When he finally did call, he said there wasn't any more work at the Chin house, and that he'd just have to get back to me when something more came up, and that was that. The end. I never heard from him again. I was figuring on something a little more formal and profound, but really, my tenure at Marshall Construction just came to a slow, stuttering, unofficial halt. One morning toward the middle of June I woke up and realized I didn't have a job any more.

A part of me wanted to be elated, wanted to feel like this was the beginning of my summer vacation and I was looking out over a whole string of carefree days stretching away as far as I could ever see. But the more insistent, more joyless part of me was filled with a slow, dark, brooding dread that I was indeed going to have to go out and find another job. As bad as Marshall Construction had been at times, it was still a known quantity, and nothing compared to the fearful images of my next construction job, which my mind kept serving up to me in quiet moments, like when I was sitting on the couch reading the baseball box scores at 10:25 on a Tuesday morning, nursing my fifth cup of coffee. I pictured a crew with five Brians and a boss with little piggy eyes and a fat gut who docked your pay every time you fucked up a cut. We'd be building a five-story tower with a 12:12 roof and it would rain all the time. Everyone would drink, and at lunch they would crack jokes about niggers and pussy and tell war stories about target practice with their automatic pistols.

Luck and the fates, however, were favoring me as spring turned to summer. Instead of going to work for another company, I began working on my own with Greg. The geodesic dome project had fallen through, but Greg and I were hired as a partnership to build an elaborate deck onto the front of a charming old house in the Waverly Park neighborhood. This led to another job, and that job led to another job, and so on. All word of mouth. Greg is an experienced cabinetmaker and trained architect, with a decidedly arty and innovative bent. He had been caught up in a personal, spiritual search for several years, and had not worked much. But he was ready now, needed the money, and was the perfect person for me to work with. We actually had fun.

When he moved back up to Washington State in the early fall, I continued working on my own, and I've been working alone ever since. Now, when the fuck-up menace emerges from its cave, there is no one to see it but me, and I become simultaneously the perpetrator and the witness, fucking up my own jobs with my own fuck-ups and paying the price for them myself.

*   *   *

Last fall, Kathleen and Emily and Shannon and I went out to the coast for the night to celebrate Emily's twenty-fifth birthday. In the morning, Kathleen and I woke up early, before the girls were awake, and went out to our favorite little hippie cafe for breakfast. We were sitting at a table on the second floor, a wonderful, isolated table with a clear view of the ocean and two big shelves of books along one wall.

I was telling Kathleen about a particularly bad fuck-up that had happened the day before. I had been working for a man named David Neufield, the father of a friend of Shannon's, who had recently purchased a house in town, which he was converting, with my help, into a duplex. It was going to be a rental property targeted for the student market. I was doing some remodeling in the old laundry room, getting the walls closed in so we could install a new washer and dryer, and stupidly, without caution or thought, I cut right through a copper water line with my Sawzall. Water coming out everywhere, mayhem, the frantic hunt for the water shut-off valve as the floor floods and the sheetrock starts to swell and crumble—it was a nightmare of epic proportions, and as I ran around trying to control the damage, I was mercilessly giving myself the same treatment Vern had so often given to the old six-foot level when a wall suddenly read half an inch out of plumb. I was ready to drive my truck and all my tools right into Rainridge Reservoir.

In the end, we got a plumber to come out and re-sweat the pipe, the water damage was minimal and easily fixed, and within a few hours you would never have known anything at all had happened. I squared all the costs with David, so there were no lingering resentments or guilt, but still, sitting there with Kathleen, twenty-four hours later, telling her the story, I could feel myself flushing again with self-loathing and humiliation.

"Joey, face it," she said. "You would never treat anyone who was working for you with anywhere near the abuse you lay on yourself." Kathleen was right, of course. What is it with me? Where does all this shit come from? "Is everyone in the trades like that?" she asked. "I mean, do you all blast away at yourselves every chance you get?" "Well," I said, "you know, mostly I think it's true, we do, although when I was working with Greg last summer we had our little shrine to St. Joseph and we said prayers and meditated and blessed the tools every morning and all of that. You know, it was a very spiritual job, and so when the fuck-ups happened, we just took it as a sign to slow down, and we'd do our invocations and laugh a lot and then proceed."

I thought about it some more, tried to remember all those jobs, all those bosses, all those fuck-ups, all those tantrums and explosions. "One other time I always remember, was when I was working for Bobby Hayward in San Francisco, when I first went to work for him, and we were in his shop putting together this huge butcher-block slab for a kitchen counter we were building. We had all these long pieces of butcher block we were gluing up with contact cement, which you just get one chance with. And that butcher block is not cheap. So we got everything all glued up, and we were ready to clamp the whole thing together, and there was not one pipe clamp left in the shop. Somebody had them all out on another job. Now if it had been me, or Sam Bazealy, or for that matter, Vern or anyone else I ever worked for, there would have been a major shit fit, I mean court marshal and summary executions all around: 'Where the fuck's the pipe clamps? Who took the fucking pipe clamps?' But Hayward just started laughing his ass off, I mean, he thought it was the funniest thing he had ever seen. When he was finished laughing, he just jury-rigged some improvised clamps with screw-down blocks and wedges, and the countertop turned out great. That really impressed me. I wish I could just plug into that instead of whipping myself with the cat-o'-nine-tails every time the slightest thing goes haywire."

Kathleen agreed. I felt better, too, after talking about it, and by my second latté I had pretty much let the whole severed water-pipe debacle fade away. I reached for an old anthology of poetry that was sitting nearby on a shelf and I said, "I'm going to open this up and whatever poem it is, that will be the poem for the day." And I opened it to a poem called "Hack and Hew" by a poet named Bliss Carman (1861–1929). The poem is about two sons of God named Hack and Hew who work their butts off creating the Universe for the Big Taskmaster and then think they can kick back and relax for a while. But God decides to make them men and send them down to earth to keep working. So that's how they end up. The last two stanzas, it seems to me, say a lot about the present human condition in general and building in particular:

> *And still the craftsman over his craft,*
> *In the vague white light of dawn,*
> *With God's calm will for his burning will,*
> *While the mounting day comes on.*
>
> *Yearning, wind swift, indolent, wild,*
> *Toils with these shadowy two—*
> *The faltering, restless hand of Hack,*
> *And the tireless hand of Hew.*

Anyway, it has been almost a year since the Chin project. I get phone calls from Brian every few months. His baby was born, a little girl named Beth Louise, and for a while he was working for Severson Construction doing big tilt-ups in the north valley, but then he got laid off. The last time I talked to him, he told me he had just gotten the certification to become a licensed asbestos-removal technician. He keeps asking me if I want to go fishing or something, but I always tell him I'm too busy, which is usually the case. I also tell him that one of these days we'll go out and play the Three Oaks golf course and take our chances on the third-hole dog-leg around Beverly's house.

Vern I've seen a few times at Central Jerome Building Supply, where I now have an account. He came up behind me at the counter early one morning a few months ago and started telling the clerk, "Watch out for this guy. Don't give this guy any credit," just joking of course, and it made me feel like I was being initiated into the brotherhood of contractors, who are always making cracks at each other about how unreliable they are and how they're all out to make a fast buck without doing any work. Still, the sight of Vern flustered me so badly I could hardly get a word out of my mouth—I felt like I had just been caught by the headmaster with a bottle of Jack Daniels in one hand and a cheatsheet for Cicero in the other. Really—it was bad. I went home that night and told Kathleen I thought I needed to go into therapy to deal with my authority problem.

Months went by, and I only saw Bud once, in the distance, in his truck, crossing through an intersection while I was a few car-lengths back. He never saw me at all. Then about a month ago, I went out to the coast with Shannon and her friend Kativa, who had come up from L.A. for spring break. Kathleen had to go off to Portland for a continuing-education conference. We got out to the motel, checked in, pulled up to our parking spot, started getting out of the car, and here comes Bud and his wife walking out of the room right next to ours. Bud and I had a long conversation standing there by our cars, just like it was the end of another day at the site. He filled me in on his news, most importantly, that Carol Anne's tumor has practically shrunk away to nothing, and that he's the father of another little girl who was born in December. He also said that Marshall Construction is getting ready to start work on a big, new development out near Alder and Hargis (fifteen lots—spec house plus pre-sold, the whole deal), and that they might well be hiring in the near future. I said, "Great, I might just give your Dad a buzz," but really, I don't think I could ever belt myself into that harness again. Still, it was good to see Bud. Neither of us mentioned Rush Limbaugh. We both agreed that it was a little bit scary to think of Brian as a licensed asbestos-removal technician, and I ended up telling him (as I have so often told him before), that I was surely going to make it out to the Speedway one of these weekends to see the funny cars and the demolition derby. On the weekends Bud is the emergency Medical Technician out there.

*   *   *

Today around noon, I drive over to a friend's house. Her car is in the shop and she asked me to take her over there across the river so she can pick it up. I said sure. So I leave her off at the Honda dealership, and then I realize I am only a few blocks away from the Chin house, and so I think, why not go over there and take a look at it in its finished state, all scars healed, all final details in place?

The day is April-gorgeous with a luxuriant sun, a sun we have barely seen at all for months, beating down on bunches of daffodils and tulips, flowering shrubs, dogwoods, and magnolias. The green that vibrates everywhere from the lawns and budding trees seems of an almost unearthly brightness, almost blinding to these rain-dimmed eyes.

In my new black truck, which I bought only two months ago, and which I will be paying for well into the next millennium, I steer down the familiar sequence of streets until I come to Cypress Way, the long, one-block cul-de-sac where the Chin house sits in the next-to-last lot on the left. I turn the corner, head down toward the house, and then see what looks exactly like Vern's truck parked out in front. It can't be. Still, my instincts flinch at the sight of that long, white Ford 250 XLT supercab with the enormous service canopy in back. If a helicopter had been flying overhead, tracking me down the street, it would have looked as if I had either been shot or suffered a massive heart attack. I steer straight into the curb, sideways to the road, and am about to back around and retreat when I think—wait a minute, why are you doing this, what are you so afraid of here? The old patriarchal authority menace. But then I think, trying to reassure myself, what are the chances of that really being Vern's truck, right here, right now, on this day, on the one day you decide to stop by the Chin house, the first day in almost a year? Bud had told me, when I ran into him out at the coast, that Vern had been called back to Beverly's once, to renail a piece of siding that had slumped away (we agreed it must have been Brian's work), but that was a one-time deal. Anyway, I get a grip, straighten out my truck, and proceed slowly down Cypress Way, still trying to convince myself that it's not Vern's truck, it's just a truck that looks a helluva lot like Vern's truck. But as I get closer and closer, as I see the familiar markings, the identical characteristics, and finally see Vern's lanky Gary Cooper frame behind the tailgate, tinkering with his tools, there can be no more doubt. Strange too is the fact that he is not parked in the driveway of the Chin house at all, but in the driveway of another brand-new house next door, a house that now sits in the vacant lot where Brian used to conduct lunch-hour golf-ball reconnaissance runs.

I pull in beside Vern's truck, get out, and come around to the back where he looks up and sees me just as I say, "What the hell are you doing here?" trying to sound casual and chatty and just like one of the guys, though the impact of this coincidence, that we should both find ourselves exactly here and exactly at the same time, feels both grotesque and at the same time almost preordained, as if by some higher mechanism of fate. I'm sure he feels it too as he comes around to shake my hand, and we laugh about it; in fact we laugh a lot, probably more laughter in ten minutes as we stand there in the driveway of this other new house than we laughed in two whole months on the job.

This new house, he says, was just built in this vacant lot, and the owners, once the barrage of golf balls began, had contacted Vern to have him put the same Plexiglas shields over their windows that he had installed on Beverly Chin's, which is why he is here today. "I only just got here about two minutes ago," he says, shaking his head at the strangeness of this synchronicity. I tell him I just happened to be in the neighborhood and thought I'd come by to see what the Chin house looked like in its final, finished form. I'm anxious to explain to him why I'm there. I don't want him to think I'm stalking him, or looking to hit him up for another job. "I mean," I say, "this is the first time I've seen it since last June." We turn together and look at the Chin house. "Is she happy with it?" I ask. "Oh yeah," he says. "You know, the usual this and that, but at least I got my money out of her."

The house looks crisp and certainly complete in every sense of the word, every board in place, every fake Mediterranean roof tile, every exterior light fixture, window screen, gutter, and scupper, right down to the brass knocker on the big, bright green front door. The driveway sits comfortably in an apron of green landscaping, with colorful spring flowers poking up out of their dark mulched beds amidst swatches of Ortho-perfect lawn. The overall effect is somewhat stark. Not like a new car, which looks its best coming off the show-room floor, and drives off the lot immediately imperfect and lower in value. A house seems to need the nicks and scratches of time to give it some sense of occupancy, to warm it up and personalize it. It is not a machine, a vehicle, or appliance—it is a home, and that is something different. This Chin house, right now, on this sunny afternoon, looks almost like an architect's model that has taken a pill and suddenly expanded out into a space it was never meant to occupy. Of course, I don't say any of this to Vern, just bob my head up and down and tell him how great it looks and all that usual crapola.

I ask him what he's up to and he says this and that and also mentions the start of the new development out at Alder and Hargis and even says, "I might be calling you up one of these days." To which I say, "Great, it would sure be fun to work with you guys again." Which could not possibly be true by any measure of the word "fun," and must seem to Vern like a real crock. In the end, he asks me about my own work and then we laugh some more and commiserate like two old veteran contractors comparing notes on remodels, common headaches, taxes, callbacks, and dry-hole client calls, where you go out to bid a job and suddenly realize you're only out there to give somebody a bunch of free advice. I tell him about all the tools I've bought and the tools I still have to buy and he says, "Oh, yeah, it's endless." He admires my new truck, which is a small Japanese truck with a squat recreational canopy that looks further diminished standing there beside Vern's big, white Ford.

Ten minutes of this is about all I can take. I already have a feeling I'm keeping him from his work; I almost have the feeling I'm back on the job and goofing off with useless chit-chat. I don't go any closer to the Chin house, I just stand there, about in the same place I was when I was sighting through the surveying instrument seventeen months ago looking at grade stakes, wild green grass, the brown earth Spike's bulldozer was curling up, and Vern's pinched face in the cold November air. We shake hands again and say "Goodbye," and Vern says he might be calling and I tell him I'll be looking forward to it. And I drive away.

Afterword   *by Catherine Sibert*

WHEN Jody was nine years old, he told God he wanted to be an architect. But by the time he was sixteen years old, he wanted to be a writer. He did become a carpenter-construction worker for many very interesting architects. He built houses and demolished them and remodeled them. But he always wrote.

Jody grew up a Boston Brahmin Catholic and went to Harvard. He was a hippie and a performance artist in San Francisco, and he wrote. He smoked a lot of pot, drank a lot and got sober in AA, and he wrote. He was a father of two daughters, and he wrote. He got a master's degree in writing, and taught writing; he was an alternative television producer and an actor for educational films; he was a devoted Red Sox fan and played ice hockey; and he wrote.

He meditated, prayed, read, went to meetings, and he talked and he talked and he talked and he wrote. He laughed and made us laugh, and he wrote: fourteen novels, a score of short stories, poetry, several screen plays, and articles.

There were no publishers. He wondered if God had stopped listening to him when he was nine, hadn't heard about the writing thing, kept sending architects, not publishers. He also had a theory that he must have been a famous writer in a past lifetime, but he had been very, very arrogant, so in this lifetime he had to write and write and write, be rejected, learn humility.

Toil was the last book he finished. About six months before he learned he had lung cancer, his agent called and asked, "Do you have anything to show me?" He had *Toil*. He sent it to her. In March, two weeks after the diagnosis, she called to say, "I love this book." He told her that he had cancer. She said, "Can you do some rewriting?" So during his last months, of course, he wrote. Jody died on June 20, 1998. Two days after the first anniversary of his death, his agent called to say, "I sold it."

If that God of his meant for him to learn humility or kindness or compassion, Jody learned it. The toil of manual work, the toil of creativity, the toil of living, and finally the daunting toil of dying was in the end done with a monk's perseverance and devotion, with stunning humor, and with love.

CHELSEA GREEN

Sustainable living has many facets. Chelsea Green's celebration of the sustainable arts has led us to publish trend-setting books about organic gardening, solar electricity and renewable energy, innovative building techniques, regenerative forestry, local and bioregional democracy, and whole foods. The company's published works, while intensely practical, are also entertaining and inspirational, demonstrating that an ecological approach to life is consistent with producing beautiful, eloquent, and useful books, videos, and audio cassettes.

For more information about Chelsea Green, or to request a free catalog, call toll-free (800) 639-4099, or write to us at P.O. Box 428, White River Junction, Vermont 05001. Visit our Web site at www.chelseagreen.com.

Chelsea Green's titles include:

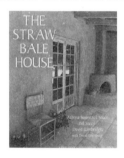 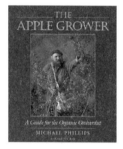 

The Straw Bale House
The New Independent Home:
  People and Houses that
  Harvest the Sun
Independent Builder:
  Designing & Building a
  House Your Own Way
The Rammed Earth House
The Passive Solar House
The Sauna
Wind Power for Home &
  Business
The Solar Living Sourcebook
A Shelter Sketchbook
Mortgage-Free!
Hammer. Nail. Wood.
Stone Circles

The Apple Grower
The Flower Farmer
Passport to Gardening
Keeping Food Fresh
The Soul of Soil
The New Organic Grower
Four-Season Harvest
Solar Gardening
Straight-Ahead Organic
The Contrary Farmer
The Contrary Farmer's
  Invitation to Gardening
Forest Gardening
Whole Foods Companion
The Bread Builder
Simple Food for the
  Good Life
The Maple Sugar Book

Believing Cassandra
Gaviotas: A Village to
  Reinvent the World
Who Owns the Sun?
Global Spin:
  The Corporate Assault
  on Environmentalism
Hemp Horizons
A Patch of Eden
A Place in the Sun
Beyond the Limits
The Man Who Planted Trees
The Northern Forest
Loving and Leaving the
  Good Life
Scott Nearing: The Making
  of a Homesteader
Wise Words for the Good Life